For permission requests or enquiries, please contact:
robethics@gmail.com

Independently published by Robethics LLC

First Edition: December 2022
Second Edition: September 2026

Interested in getting involved?
Join the Responsible AI community at robethics.org

*To my wife Emily,*
*our children Julia and Lucy,*
*and the 100% trustworthy algorithm*
*that writes my dedications*

# Table of Contents

**Part Three: Emerging Risks**



**Part Four: Call to Action**

# Preface

The innovation brought by Artificial Intelligence (AI) is set to change our society and our lives as never before. AI is already transforming a number of domains such as education, finance, transportation, defense, energy, and healthcare introducing extraordinary new opportunities but also fundamental questions about the ethical implications of the use of AI and the type of society we want to live in. In order to be ready for this new world, we need to ensure everyone is meaningfully included in the debate.

While technical topics related to benefits, risks, and ethical implications of deploying AI are discussed within small circles, mostly among engineers and data scientists, there is still yet to be a meaningful involvement of the general public. This is an enormous gap which needs to be addressed. AI ethics, specifically, while not a new discipline, is a neglected area in the public debate. Today's landscape lacks, in fact, a comprehensive view and proper categorization of all the different ethical issues that intersect with emerging technologies such as AI. This is not an easy task, as the challenges we face with AI are complex and intrinsically multi-disciplinary. As such, they require expertise and insights across a wide range of disciplines: computer science, data science, philosophy, statistics, psychology, law, sociology, risk management, international relations, public policy, economics, business management, and everything in between.

My personal goal is to help expand the conversation on AI to as many citizens as possible. There are too many unanswered questions around the risk and impact of AI to let it go unmanaged. I want to make this topic become a dinner table discussion accessible by everyone so that we can start finding those answers together, as a community, rather than letting a few tech companies (and those who control them) decide for us. While I do not consider myself an expert in any of the many fields that intersect with AI, it is true that my personal multi-disciplinary background puts me in a unique position to ask meaningful questions and even help connect the dots.

The objective of this book is to deliver what is missing today: a comprehensive overview and categorization of risks, opportunities, and ethical implications of the rise of intelligent machines methodically organized in buckets based on the different types of technologies and applications. It also provides practical action items for concerned citizens, activists, and firms to ensure a safe and beneficial AI. While this work will hopefully be appreciated by AI professionals and other technical experts, it is intended to be easily approachable by the general public, including those exploring such complex topics for the first time, and provide a myriad of case studies and examples to which everyone can relate even if new to this field.

The book is structured in four sections:

Part One focuses on the latest technologies and related applications. It first provides an analysis of the most significant technological advancements seen in computer science, highlighting key drivers, functionalities, and technical limitations. It then covers a number of ethical questions with regard to the increasingly relevant role played by machines in our society, with a closer look at controversial topics such as machines' self-awareness and moral agency. It finally looks at how such technologies are being applied today across functions and sectors highlighting risks and opportunities.

Part Two focuses on the ethical questions related to AI, common misconceptions and the biggest obstacles to ethical AI. It then provides an overview of unexplored territories and a deep dive into five specific case studies: self-driving cars, personal robots, lethal autonomous weapons, virtual worlds, generative AI, and autonomous AI agents.

Part Three focuses on the most relevant risks we are facing with AI. It starts with risks related to AI safety, quality, fairness, privacy, and transparency, and then moves to more fundamental risks such as value misalignment, inadequate human oversight, labor displacement, environmental impact, impact on democracy, and existential risk.

Part Four provides a framework to navigate the ethical assessments of intelligent machines across the lifecycle of products and technologies. It also calls for a call to action with a road map for citizens and business leaders to ensure an ethical AI. The proposed solutions range from national and international cooperation to education, research, and diversity. It also includes a governance toolkit for firms interested in actively managing their AI pipeline to ensure ethical outcomes throughout its lifecycle.

Is AI the best thing that has ever happened to us or the worst? The most likely scenario is that it will be both.

# Part One
# **AI Among Us**

# Introduction:
# Our World will Bever Be the Same

In 2016, Google's Deep Mind team challenged Lee Sedol, the 18-time world champion and undefeated master of the board game GO. The game GO, originally from China, has been around for at least 3,000 years and it is considered the oldest board game continuously played to the present day. It is played by 40 million people worldwide. It is a bit like chess, but it has many more squares, and its black and white pieces all have the same value. The game is complex, much more than chess as each turn has about 200 different possible moves compared to just 20 moves available with chess. It is also a constructive game, where one piece is added at each turn so the number of possible variations of all positions is so high that it is even higher than the number of all the atoms that exist in the observable universe. It is impossible to handcraft a set of rules to help make the best move. The only way to play is to use intuition and creativity, rather than memory or calculations.

Who won? AlphaGo defeated Lee Sedol 4 to 1. What mattered was not just the fact that the machine won, but also that it won thanks to an unexpected (creative?) winning move that no one ever saw before: with the now famous move 37 from game 2, AlphaGo surprised everyone by putting a piece on the 5th line, something that no professional would ever do at the beginning of a game, which put AlphaGo in a position of strength.

This was not the first time that machines won against humans. One famous example from the nineties was when IBM's Deep Blue computer defeated Chess Grandmaster Kasparov. The computer was given an inventory of all possible moves and instructed to select the best move among every possible one. As in the kids' game Tic-Tac-Toe, where each scenario always has a best move, the programmers mapped all possible moves and identified for each scenario the move with the highest likelihood of winning. With the game GO, however, there are so many scenarios that it is impossible to map them. No one gave AlphaGo a pre-programmed set of instructions on winning moves. Developers only showed the machine a number of experts' moves, and then let the machine play against itself millions of times. The machine then independently learned from its own mistakes and autonomously figured out how to play and how to win. This is an important detail because it demonstrated that not only can machines select the best move from a massive archive of past moves, but they also can creatively come up with new ones never seen before in the history of the game.

AlphaGo's victory against Lee Sedol was just the beginning. What happened right after was even more incredible: a new version of AlphaGo played 60 games against all top 20 players in the world and won against all of them, coming up with many more breakthroughs, instantly overturning 3000 years of wisdom in this ancient game. Within less than a year, Google created a brand-new system with zero knowledge on the game and zero examples of past games, called AlphaGo Zero. AlphaGo Zero instantly beat AlphaGo and did it again 100 times and won 100 games to none. What does it all mean for us?

First, it confirms that AI can outperform humans even for tasks learned without instructions. Can you imagine a newborn that within hours is able to finish every video game? And all of that without any notion of motion or time, and without any instruction on how to play or even the rules of the game? For AI, knowing that there is an image on the screen and that there is a score to beat is sufficient to learn how to play at superhuman level. And if machines are so much better than humans at such a complex game, they are going to be better than us at activities that are easier than GO, which amounts to a lot of different things. As machines continue to outsmart us at more and more tasks, how is that going to affect our society and our role in the world? How will machines affect our labor market? With machines able to easily replace humans not only for low-complexity repetitive jobs but also for highly complex tasks, virtually all professions will be impacted, including those considered immune from this revolution such as medical doctors or lawyers. Are we all going to be replaced by machines? How is the transition to this new world going to play out for our communities?

Second, from a more philosophical perspective, the more we explore this topic the more the discussion becomes less about algorithms and robots and more about ourselves and what it means to be human. We always thought intelligence set living beings apart from everything else. However, if the most complex things we do can be performed by a machine, our role as "the intelligent ones" necessarily comes into question. Across history, humanity has repeatedly been forced to reconsider its place in the world. Copernicus showed that Earth is not the center of the universe, overturning a worldview in which we were at the center of everything. Centuries later, Darwin demonstrated that humans are not separate from the rest of the living world but part of the same evolutionary process that shaped all animals. With Freud, human consciousness lost its centrality as was revealed to be only a small part of a much larger psychological machinery driven by unconscious forces. Today, the rise of artificial intelligence marks another profound shift. Tasks that were long considered uniquely human as they required intelligence, are increasingly performed by machines, challenging the idea that the ability to carry out activities that require intelligence is exclusive to humans. Does our brain possess unique properties that would make our intelligence impossible to fully replicate? Or is it time to hand the torch of intelligence to the machines we have created so that they take it over from here, as a baton in a relay race that goes beyond humanity? If the ability to complete complex tasks that would require intelligence is no longer the differentiator, then what makes us different or special? Will we soon lose the exclusive right to use "sapiens" as a qualifier of our characteristics?

The exponential growth of technological advancement puts us at the beginning of a new era. Every few generations there is a world changing technological revolution. AI is one of them. After the three industrial revolutions that saw the advent of steam engines, electricity, and digital technologies, we are at the beginning of a fourth industrial revolution that will fundamentally change our civilization and change our society much faster than the previous revolutions. We have built machines to help humans with physical work, and then computers to help with calculations based on humans' specific directions. With AI's learning capabilities, we have machines that require no instructions, only goals. The change is so big this is widely considered the 4th industrial revolution. The opportunities are unprecedented, but so are the risks.

On one hand AI has the potential to help humanity solve its biggest problems, from social inequality, to curing deadly diseases, to climate change. In some areas it will free us from doing things we do not want to do, as we consider them boring, tiring, or dangerous, in others it will allow us to improve how we do things and even help us accomplish tasks we have never accomplished before.

On the other hand, AI can easily lead to unintended consequences, including unethical outcomes such as violations of privacy, racial discrimination, and social manipulation. Some experts now believe AI could put democracy at risk and in the long run, even trigger much more catastrophic events, such as the extinction of the human race. In fact, given the risks associated with AI, some have already started referring to AI as our humanity's final invention.

We do not know how, and how quickly AI will evolve but we know we are heading toward a world with an increasingly powerful technology. With the invention of the steam engine, we went beyond humans' physical force. The activities performed by humans progressively became less about physical work and more about using our cognitive abilities to make decisions. With AI we are going beyond human thinking abilities to a point that we have already started delegating to machines highly complex tasks. We now even have AI software able to design new types of AI: AI that develops AI!

Are we ready for this? How can we mitigate so many risks? How can we prevent such an unprecedented power from getting in the wrong hands or out of human control? What can we do to ensure that AI will be used ethically and that it will benefit all of humanity rather than putting it at risk? How can we be in the driver's seat and not be caught by surprise by AI's unintended consequences? And what if we combine the power of AI with the physical strength of robots?

Two years after the eye-opening AlphaGo victory, another very relevant moment for me happened when Boston Dynamics released a video demonstrating how their big robo-dog, Spot, successfully managed to open a door and go to a different room despite a person attempting to stop it. After seeing the power of AI embodied in a physical form resembling the shape of a familiar living being, the philosophical questions around intelligent machines took a very tangible physical form. On one hand it is difficult not to be excited about its deployment in dangerous activities such as construction, mining operations, or monitoring of highly hazardous areas like the radioactive sections of power plants. Even deep ocean and space exploration would greatly benefit from the technology allowing machines to reach places that no humans could directly experience. On the other hand, it is very easy to imagine scenarios in which such machines do not operate as intended, get out of control, or end up in the wrong hands. Given technology's evident link to military applications it is not hard to believe at some point similar machines would be used by either law enforcement or criminals in communities around the world leveraging not only physical force, but also an ability to scan faces, record conversations, etc. Do we want to have these machines around at all? What activities should they be used for and what activities should they be never used for? Considering the robots' power, sensors, radars, thermal cameras, and the AI enabling them to operate independently and even foresee our moves, how do we ensure this type of technology will always be safe? Will there always be an easy to reach big red button to shut them down? What is the line that a robot should never cross?

More recently, with the release in 2022 of OpenAI's chatbot ChatGPT, the power of AI has suddenly been made available to millions of people around the world with its stunning ability to generate language in a very creative way, opening the door to new fascinating opportunities but also new risks and fundamental questions about the future of our society and of human culture itself. What will it mean for us to live in a world where a significant component of text (from novels to legal opinions) and all other types of content such as images and music are created not by humans, but by AI? How will it affect the way we learn, think, communicate, and participate in the democratic process?

Ultimately, how can we ensure that our civilization will use AI's power so that it will in fact become the best thing that has ever happened to us rather than the opposite? Let's discuss.

# The Power of Machine Learning

## What is Artificial Intelligence?

The term Artificial Intelligence does not have a generally accepted definition, nor a set of minimum requirements that could be used to identify if a technology or a methodology qualifies as such. This is primarily due to the fact that the term is used differently across a variety of disciplines and functions, such as computer science, philosophy, neuroscience, public policy, etc. It is also due to a lack of clear definitions of the term "artificial" and the term "intelligence". The use of the word "artificial" to describe machines' intelligence is fairly straightforward: at a high level, it is called "artificial" because it is the outcome of an artifact, a man-made object. What has triggered the most heated debates is the term "intelligence". The divergent views around such a term are mostly due to the fact that it is an area in which we still know very little. While intelligence is undoubtedly one of the most exciting products of the universe, it is something that is still not clearly defined and understood.

In addition to not having a definition, AI is a field where there tends to be a lot of confusion among the general public. For example, thanks to Sci-Fi novels, Hollywood movies, and low-quality journalism, it is not uncommon to immediately associate AI with shiny humanoid robot with an athletic body, with either an evil-looking face like the main character from the Terminator movie, or a peaceful angelic expression perhaps performing hilarious activities for a robot such as typing on a computer keyboard or trying to touch a hologram. The term AI is also often loosely used to describe any type of not-yet-achieved future technology, to a point that it has become synonymous with anything that has not been invented yet.

From an historical perspective, the term Artificial Intelligence was coined in 1956 by the authors of a proposal for a seminar to be held at Dartmouth College to study techniques intended to simulate humans' cognitive abilities. Since then, AI has had a number of periods of hype, and other moments of reduced interest, called AI winters.

In 2010, AI has experienced a new boom, and the term has been adopted by a number of disciplines and functions, typically to describe any type of technology that is able to mimic human capabilities or perform cognitive functions we generally associate with human minds, such as, reasoning, learning, problem solving, interacting with an environment and, more recently, even exercising creativity. The drivers of such a boom have been considerable improvements in machines' processing power, access to massive quantities of data, and the advancements in a specific type of methodology called Machine Learning (ML).

## AI as Machine Learning

In Computer Science, AI is often used as a synonym of ML, the pattern recognition technology behind the most promising and potentially disruptive aspect of AI.

During the first decades of AI, models were merely deterministic, aimed at processing information based on rules built in the programs. Only at a later time computer scientists came up with probabilistic ML models where some of the program's rules are defined by the program itself. At the risk of oversimplifying what ML is, it is a software that learns by examples. Rather than being programmed using deterministic rules (if A happens then do B), the machine autonomously finds its own set of statistical rules and it is, therefore, able to answer a question without being specifically programmed for it. The mathematical equation designed by the data scientists or software developers, also known as an algorithm, looks at the data and recognizes patterns or rules that allow it to create a model that can be used to make predictions.

This is not very different from the way animals learn. For example, if we want to teach a dog to sit when we say "sit", we do not need to program the dog's brain. The dog's brain writes that code itself based on our consistent positive reinforcement whenever the dog happens to do what we want him to do. Similarly, with ML rather than providing a specific set of instructions on how to produce an output, we just show a machine what a successful output looks like and let the machine identify the best set of rules to get to it. For example, it is thanks to ML that a real estate app is able to accurately guess the future sale price of a house. By looking at multiple data points related to a specific market (square footage, yard size, location, school ratings, etc.), it is then able to create its own statistical model and use it to produce a price forecast. Another common example of ML is a model used by online shopping and streaming sites to make customized recommendations based on personal profiles, preferences, etc.

Thanks to this ability to self-program, AI can learn how to solve complex problems in creative and unexpected ways. In addition to the famous Go breakthrough, there have been a number of cases in which ML has produced incredible outcomes. For example, an algorithm tasked with learning how to play the Atari Brick Breaker video game, figured out very quickly that the best way to win was to direct the ball toward the top of the screen to build a tunnel above the bricks. Even more interestingly, in 2013 an algorithm playing Tetris decided to pause the game indefinitely to stop the blocks from filling up the game area, therefore reaching the goal of not losing the game! Was that one of the first examples of AI cheating or one of the first clues of a value misalignment that will profoundly affect our relationship with machines?

Regardless, it is impossible not to think of the famous quote from the 1983 cold war movie War Games, in which the computer independently came to the conclusion that "the only winning move is not to play".

## Neural Networks and Deep Learning

The most advanced version of ML is based on a technology called Neural Networks. This sophisticated evolution of ML uses a layered structure of algorithms called an artificial neural network, the design of which is inspired by the biological neural network of the human brain. Those networks are built in layers that independently discover patterns and identify rules using very complex statistical models. The data goes into the system, it is analyzed, reorganized through different layers, and then finally used to produce an outcome such as a recommendation or a score. This allows machines to solve very complex tasks such as recognizing patterns, categorizing images, sounds, etc. For instance, an application can be trained to recognize a stop sign by giving it many pictures of different stop signs. The first layer of the model learns to figure out the shape, the next one the color, until the machine comes up with its own set of rules.

Given the computer power needed to analyze so much data, the first generation of ML struggled to go beyond just a few layers. Thanks to the recent advancements in computer hardware, Neural Networks can now easily deploy many more layers. In fact, it is not uncommon to have models with hundreds or even thousands of layers as long as there is sufficient data to train the machine. Just to put this in perspective, models need about 15 layers to distinguish a dog from a cat. With 100 layers visual recognition can distinguish objects across thousands of categories with human-level accuracy. When models are able to leverage more than just a few layers they are generally referred to as Deep Learning algorithms. Given the complexity of fine tuning artificial neural networks, there are now optimization algorithms built specially to build better algorithms: AI building AI. What is important from an ethical perspective is that many of the layers might not be fully accessible by humans. We generally have access to only two of the layers: the input layer and the output layer. All the other layers in between are hidden so it is very difficult if not impossible for humans to see if the data is processed in unethical ways. This creates "black box" situations where no one is able to assess what takes place in the intermediate layers of the models being used.

There are two main types of machine learning methodologies. In certain models, called Supervised, the data fed to the machine is already labeled by the developers. For instance, if the machine is learning to distinguish a dog from a cat, the developers feed the machine with pictures called dog or cat, so that the algorithm can create its own statistical model to correctly identify which animal it is based on all the labeled images used for training purposes. The model independently identifies all the key similarities across all the images labeled as cats and all the similarities across all of the images labeled as dogs. When the model is presented with a completely new (unlabeled) image, it compares it against the common patterns for cats and for dogs to find the most similar. In the end, Supervised Machine Learning is very similar to the way a parent teaches a child what cats and dogs look like, pointing at them in a book and verbally labeling them as dogs or as cats. AI learns by looking at examples. With enough examples, it extrapolates patterns related to all kinds of groups or categories.

In other models, called Unsupervised, machines independently discover patterns via clusters or more complex techniques. For example, a machine can be trained to identify the criteria associated with fraudulent financial transactions just by comparing a number of data points such as location, amount, etc. In one version of it, called Reinforced Learning, the system attempts a variety of actions based on trial-and-error techniques. At each attempt, it then receives feedback (typically as a score) on the quality of its output which allows it to improve the model and ultimately reach the goal. The whole process is very similar to the techniques used to train pets: rather than explaining the intended behavior, the trainer gives them an objective and provides rewards every time they get it right. The trainer is not fully able to understand how it is all coded in the pet's brain but through repetition the pet develops its own model. The correlations between data points that machines are able to identify are so complex that they are sometimes not immediately understandable by humans and sometimes look irrational or arbitrary. An interesting case, for example, emerged at a hospital using AI to prioritize patients in the emergency room. The doctors noticed that the algorithm tended to de-prioritize people showing up with pulmonary issues (suspected pneumonia) if they also had a history of asthma. While it was initially considered an error as evidently counterintuitive, they soon realized that the logic was actually correct as people with asthma tend to go to the hospital earlier than other groups and therefore are less in need of urgent attention.

## Key applications

Thanks to the advancements in ML, AI has proven to be very promising in a number of areas.

Forecasting is the field where AI's true power is. The more data AI is fed, the more accurate the forecast. Even a small increase in accuracy can have a substantial impact on error rates: if the accuracy is increased by five percentage points from 85% to 90% it means you cut errors by 1/3, but if increased by about 2 percentage points from 98% to 99.9% it means you cut errors by a factor of 20. These predicting abilities are the core of the many AI applications we see every day. For example, social media applications use information about your profile and your online activity to make predictions on what content is most likely to increase your face time on the platform and generate more advertising revenue. With similar types of models, ML can predict which movie a person is likely to want to watch next, which word a person is about to type in order to autocomplete a sentence, or whether an email should be moved to the spam folder. A similar approach is used by the algorithms that determine what we see when we run a search on Google. The algorithm predicts what we are most likely to find interesting by comparing keywords against a number of data points using searches made by other users, our own search, our demographics, location, etc. Those capabilities have revolutionized marketing strategies and are enabling new business models. For example, online shopping sites know so much about our needs they are able to comfortably predict and pre-order items on our behalf: from shop and then ship, to ship and then shop. This predictive ability is one of the areas making AI so similar to some of the components of human intelligence. From an evolutionary perspective, the newest area of our brain is dedicated to making predictions. This ability has driven our success as a species, and now, the success of machines.

One of the key applications of AI's prediction abilities is the development of risk scores used in the medical field, human resources management, law enforcement, criminal justice, etc. The use of such scores has relevant ethical implications especially when based on inscrutable neural networks as decisions affecting humans' lives are put in the hands of machines without the opportunity for humans to apply their own judgment to validate the algorithm recommendation or score. At the same time, the use of AI-driven scores could contribute to identifying situations in which decisions made by humans are influenced by subjectivity caused by external noise. For example, in the field of criminal justice, AI could help flag sentences where very similar cases led to different decisions based on non-relevant items such as the time of the day or the day of the week of the court hearing, influencing perhaps the mood of the court.

Other promising applications are in the field of computer vision and voice recognition. Computer vision is the term used to describe machines' ability to capture, analyze and recognize digital images. One growing subset of computer vision is face recognition software, used mostly in the law enforcement sector. As we will see later in the book, an improper use of such technology may lead to a number of risks such as bias and invasion of privacy. Another significant application is the object recognition software used by autonomous vehicles to recognize aspects of their surroundings such as roads, pedestrians, street signs, etc. Material risks associated with image recognition have already emerged across different areas. For example, there have been cases in which the software used by self-driving cars has been unable to recognize people of color due to lack of sufficient training data across all skin colors. Computer vision failures have led to a number of lethal accidents such as the 2016 crash in which a self-driving car crashed into a semi-truck as allegedly the autopilot's sensors were not able to distinguish the truck from the surrounding environment. Other important safety issues might arise from the fact that objects might have invisible markers, placed to intentionally trick the computer vision software causing an incorrect categorization of the object. In a recent test, a 3D-printed turtle was intentionally designed with invisible markers so that the AI software would see it as a rifle, not a turtle. What if a malicious person intentionally vandalizes a stop sign on the street with invisible markers so that self-driving cars recognize it as a tree?

Additional applications of ML include speech recognition and large language models (LLMs). Speech recognition allows machines to interpret in real time the meaning of verbal instructions or questions. Thanks to the many virtual assistants in our homes, cars, and wearable devices, speech recognition is already part of our daily life. Other types of ML, called language models, allow machines to interact with us by speaking our language. Combining those together makes communication with the machine very natural and intuitive, giving us the impression of conversing with a real person. The reality is that there is no person on the other side, just a statistical model combining words from enormous datasets with words and sentences from the past.

## The AI Ecosystem

The magic of machine learning is the outcome of the favorable convergence of a number of factors that created a perfect ecosystem.

First, the exponential increase of machines' computing power capabilities has enabled computers to process vast amounts of data, to a point that hardware constraints are no longer considered an issue. With cloud computing, such power has become widely available allowing businesses to safely store, process, and access data anywhere. Thanks to the recent breakthrough of quantum computing and quantum neural networks, AI will become much more powerful and able to create highly complex models in seconds rather than hours, opening the door to new applications, from cybersecurity to machine-brain interfaces.

Second, the amount of available data (and the ability to store it) has grown to unprecedented levels. The growth has been exponential: 99% of all data in the history of humanity was produced just in the last 2 decades. And this trend continues to accelerate: throughout our everyday life, whether we are spending time online while at home or running outside using wearable devices, an incredible amount of data is recorded and analyzed to understand and predict our behaviors, finding correlations across countless data points. Data is the lifeblood of AI because more data means better predictions. Going back to the AlphaGo example, the model used for the first set of games was trained on 30 million different games, more than any human would play in a lifetime. AI is constantly hungry for data. A huge unregulated market exists where companies are trading data about many different aspects of our lives. The types of data that companies can easily obtain include anything from spending habits, whether we pay our bills online, if we buy plus-size clothing, magazine subscriptions, or whether we ride bicycles. Data has become such a precious commodity that many refer to data as "the new oil". While it is true that data is very valuable, the comparison is however not particularly appropriate as data and oil do not have much else in common. For example, while oil is a finite resource that can be used only once, data is reusable and virtually permanent. Unlike oil, data is also exponentially increasing and easy to duplicate, transport, and share.

The fact that algorithms need so much data is surely a big issue for AI. Just for learning something as simple as recognizing a cat, it requires tens of thousands of different pictures of cats, when a 2-year-old child only needs to see one a few times. As algorithms continue to improve, there will likely be a strong focus on reducing the amount of data needed. After so much emphasis on "big data", there will probably be more and more emphasis on "small data" as the next big thing.

Third, the relatively small size and the strong connectivity capabilities of the machines has resulted in an unprecedented ubiquity of intelligent machines. From webcams in our cities' streets, to smart appliances, to medical wearables and implants that monitor physical conditions, interconnected machines are now everywhere. This phenomenon, often referred to as the Internet of Things (IoT), has fueled an even faster growth of available data and real time processing of information via cloud technology. Given that the number of people with access to online networks is already much smaller than the number of connected devices, we have already reached a point in which the communication from one machine to another machine is more extensive than the communication between humans and machines.

Fourth, substantial improvements in software-hardware integration have allowed us to further bring the power of AI into the physical world via robots. While we still lack a commonly accepted definition of the word robot, it is generally used to describe machines that are able to sense elements of their environment, process the information, and take some type of physical action. Based on this description, something as simple as a thermostat or a dishwasher can meet the definition of being a robot, but where things start to get interesting from an ethical perspective is when those machines, thanks to AI, go beyond purely deterministic behaviors and start behaving with some level of autonomy. Thanks to this integration we now have machines that are physically strong, have sensors (sound, vision, thermal, radars, etc.), can communicate with each other and with humans, can operate independently, and even monitor and foresee our moves.

Robots are already everywhere, and they are here to stay. Their presence is increasing not only in dangerous environments such as construction sites, mining operations, deep ocean and space exploration, nuclear plants, and the military, but also in virtually every other environment, from schools, hospitals, and restaurants, to factories, construction sites, and warehouses. Their superior physical strength and speed, coupled with some level of unpredictability has already caused many to raise concerns around their safety when working side by side with humans. Given their ability to collect and analyze data and their connectivity with computer networks, serious questions have been raised around security and privacy implications with regard to the data collected by those machines.

As a society, we will soon need to assess what types of robots we want to have around us. For example, what are the types of roles robots should have? What are the problems that we want them to solve? What do we want them not to do? Even more importantly, how do we ensure that the deployment of robots ultimately leads people to live better and not worse lives?

As technology continues to evolve, the delineation between humans and machines becomes less straightforward. Innovation has started to question the boundaries between organisms (biological, natural) and inorganic artifacts (mechanical, artificial). There will be more and more technology in the biological world while technology is adopting more biological components. Rather than a clear delineation there will be multiple gradual variations.

On one hand there will be more situations in which part of the human body is replaced or augmented by machines. From limbs to organs, to the brain itself, more machines will be part of our physical body. For example, a good portion of today's research, such as the work done by Neuralink, is aiming to connect the human brain to computer sensors to allow direct communication between neural activity and machines. Will the Internet of Things (IoT) be followed by the Internet of Bodies (IoB)? This might lead to unchartered territories where humans are so greatly enhanced by artificial machines such as prosthetics or implants that the non-biological component becomes predominant. It is impossible not to think of the Hollywood movie Robocop where a person's brain is implanted into a humanoid robot. Will a person always be considered as such as long as there is some live biological material even if most of the body is non-biological? Or would it be just a machine? We might even get to a point in which not even the person's brain is in a biological form, as it is reproduced in a super-server. Some experts have predicted in fact that one day machines will offer the opportunity to expand the computing and memory capacity of our brain. A few of them have even hypothesized that one day, that technology will be so advanced it will allow a full download into a machine of all the data contained in our brain, allowing our personality and all our memories to live beyond the death of our physical body. Some others have even hypothesized that human biology itself will be replaced with non-organic matter bringing humans much closer to supernatural beings.

On the other hand, there will be situations in which artificial machines will be made of biological material. At a microscopic scale, we already have a preview of such technology with micro-robots designed to treat diseases at the molecular level, as for example, with the microscopic robots called xenobots. While they are made with organic cells, their DNA is altered so they can be programmed to complete certain tasks within human organs. For instance, the robots can be injected directly in the bloodstream to complete specific tasks. They can even be programmed to replicate. What if a robot is made of artificial organic matter rather than inorganic matter such as in the famous 1920 science fiction play R.U.R.? Is the fact that an object is made of organic matter relevant when we assess whether it is to be considered just a machine or something more?

## Machine Unlearning

An emerging concern about the training of algorithms is around the complexities of removing ("unlearning") from AI models the impact of specific pieces of information in response to ethical and legal issues such as bias-related concerns, outdated data, or privacy issues. Once a specific piece of data is used to train a model, removing its influence to make the algorithm forget the set and unlearn the properties gained from the examples to be forgotten comes with enormous challenges. One option is to just remove the unwanted dataset and fully retrain the model. This approach, while safe, typically requires extensive computational resources. Another option is to retain the model and just make proper adjustments to eliminate the influence from the outdated, sensitive, or inaccurate data by altering its weights to resemble a model that would have not learned from such data. This is however technically very difficult, and the AI community has only now started exploring how to efficiently remove specific data influences while preserving the overall model performance. Even from a policy and governance perspective it will be key to ensure that the altering of how certain data points influence a model will not be done for unethical purposes causing, for example, an increase of deceiving or biased models.

## Artificial Unintelligence

Despite AI's learning abilities, we are still very far from achieving any type of machine understanding abilities, to a point that there are entire publications on the topic. There are, in fact, very fundamental and highly interrelated weaknesses that need to be considered as new solutions are designed and deployed.

First, AI lacks the ability to understand basic principles that we consider obvious such as the fact that objects are tridimensional and still exist even if they are temporarily not visible, the existence of the force of gravity, or the difference between correlation (A statistically associated with B) and causation (A causing B). While machines are excellent at finding correlations between data points, they are in fact unable to figure out if there is a cause-and-effect relationship between them. For instance, machines can easily find a correlation between a sunrise and a rooster singing, but they are unable to identify if the singing is the cause or the effect of the sun rising.

Second, AI lacks the ability to understand common sense. No matter how good statistical models are, they cannot alone deliver understanding of the real world. While a single machine might outperform all humans combined when asked to solve arithmetic calculations, the same machine has no idea whether one apple is better than two. The gaps are particularly evident when AI is asked to analyze text. Even for elementary sentences, if there are missing segments or there is some level of ambiguity, AI has no clue on their meaning. For instance, if we say that "the trophy doesn't fit in the suitcase because it's too large" or we say "the trophy doesn't fit in the suitcase because it's too small" we immediately know that in both sentences the suitcase was too small and the trophy too big. A machine will not know that unless it learns by observing all the countless specific rules via a data-heavy training process such as with an unsupervised learning model. The fact that common sense is not so common, as stated by Voltaire 300 years ago, continues nonetheless to be true in the world of AI.

AI's inability to understand common sense has already resulted in noticeable failures. In 2022, a Tesla that had been parked at an airport crashed into a private jet while being summoned by its owner, as allegedly the model did not recognize the jet as an obstacle due to lack of training. Some other examples of unintended outcomes have involved the social media algorithms tasked with censoring adult material. For example, a campaign to raise awareness about breast cancer was quickly shut down by the Facebook algorithm as allegedly the model deemed the campaign to be pornographic material. In a similar case, Facebook mistakenly censured the iconic photo of the Vietnamese girl escaping from napalm bombs after her clothes were ripped off by the explosions.

In another interesting case, a face-recognition system used in Asia to identify and send fines to people crossing the street outside the crosswalk lane, ended up issuing a fine to a popular figure whose pictures were on a bus riding along the street. Once again, something obvious for humans is not as clear for AI. There is not an easy solution for this as no matter how much more data we feed the model to address specific scenarios, there is currently no way to address the root cause of the problem.

Third, when used for image recognition, AI is unable to differentiate relevant components from non-relevant ones. It is not a coincidence that the most effective way to test whether an attempted website login is made by a person or by a computer, is to ask the user to recognize specific components of simple pictures through the widely used “I am not a robot” tool. Passing the test, typically simple for a human, is nearly impossible for the best machine around as understanding the relevance of different aspects of an image is in fact very challenging for all types of AI. For example, if a machine is trained to distinguish a wolf from a dog by just looking at pictures it might easily end up creating a model that is primarily based on background characteristics (whether the animal is in a forest or indoors) rather than the animal’s features (color, shape, etc.). Recent research showed in fact a similar type of error in which a dog was incorrectly classified as a wolf due to the fact that there was snow in the background, as the training data for the dogs’ images were predominantly presenting dogs inside or in warm climates. In the medical field a similar error was discovered in software intended to help diagnose malignant moles. As rulers were frequently placed next to malignant moles, the AI model incorrectly identified all moles without a ruler next to them as normal. Similarly, a promising phone app designed to identify patients infected with Covid 19 just by looking at them received an initial enthusiastic response, but it was later dismissed as further tests uncovered that the algorithm was placing heavy weight on whether the patient was laying down rather than standing up, as the images of the patients that were actually sick used in the training data showed them laying down.

Fourth, AI tends to perform poorly in any type of unpredictable or open environment. For example, algorithms used in self-driving cars are excellent in predictable environments such as low-traffic highways with perfect visibility, but much worse in highly variable settings such as high-traffic areas with bad weather or unusual driving behaviors. While AI might be unbeatable in very contained settings such as a chess game where the only rules are the rules of the game and the only variables are the game pieces, its power does not easily translate into the same type of power in a real-world environment such as, for example, a soccer game, where following the rules is just a small component of the task.

Given all those limitations, there is still a long way for machines to go to get closer to human-level abilities. While machine learning goes beyond our capabilities with regard to pattern recognition, this ability is still very limited to just the specific applications and contexts for which those machines were programmed, and work dependably only within the scenarios anticipated by their programmers. Beyond those very limited contexts, AI is too unintelligent to be trusted. Given its inability to understand and navigate the world, when used beyond the specific context for which it was designed, AI is nothing more than a clever idiot that will inevitably surprise us with absurd and sometimes dangerous miscalculations such as the ones seen above. Those who have a brain and know how to use it should not worry about the risk of being replaced by a machine because in a world in which activities are executed with AI, human intelligence will be more needed than ever.

## Managing AI's limitations

As we have seen, there are tasks where AI performs well and tasks where it performs poorly. AI is very good at making predictions in very stable and controlled environments such as playing the board game GO, and it is not good at navigating contexts with high levels of unpredictability and ambiguity or with noisy and incomplete datasets. It is also true that humans are much better than machines when dealing with uncertainty, but not as good when dealing with large sets of data.

The main trend so far has been to blindly try to replicate human intelligence whenever possible and to automate all kinds of tasks, even those that people enjoy and perform very well. Instead, we should be more selective and strategic in the way we channel our resources dedicated to the development of AI so that we maximize its benefits. Although both human and machine intelligence have limitations, they are also somewhat complementary, so the key is to assign the right type of intelligence to the right type of task so that we do less of what we dislike, or we are not good at and focus on what we do best or enjoy doing ourselves. For example, if I had to choose between spending my evening washing dishes or reading to my kids and I could delegate one of the two tasks to a machine, I would surely delegate dish washing to a machine, and spend time reading to my kids because it would be more pleasant for my kids and more rewarding for myself. It should not matter that the dishwasher performs that task in a way that is different from the way I would. What matters is that it does it at least as well as I would. In fact, it might very well be more efficient, effective, and even more environmentally friendly than me, so that the advantages of delegating the task to a machine could be multiple. Similarly, when we think of autonomous driving, we need to assess in what type of scenario (road, traffic conditions, weather, etc.) it makes more sense to let the algorithm drive and in which conditions humans should take control, rather than delegating all tasks to the machine. Because of these types of considerations many have started describing AI as a tool to augment, not to replace, human intelligence.

Considering the limitations of AI and the need to keep it within predictable environments, it is going to be important to see how much of our physical world will need to be adapted (vs. the other way around) so that AI can be used safely. This analysis will necessarily need to include a methodical risk assessment and impact analysis for the changes that will be introduced in our environment. For example, to what extent would we be willing to rebuild our highway system if that made it easier for self-driving cars to navigate them? The success of autonomous shuttles in service in many airports around the world is due primarily in part to the fact that they have been developed in controlled environments created exactly for those types of shuttles. The same technology would not work as well in other contexts unless the physical infrastructure is altered to address the shuttles' technical limitations. This is important, as very much is tied to the larger question of whether machines will adapt to humans or if humans will adapt to machines. It has happened already with many technologies in a very big way. If we look at cars again, now that we need them for everything, we might see them as a solution, but it is also true that we need them for everything because we built our cities around cars (suburban neighborhoods, office parks, street malls, etc.) rather than humans. Consciously or unconsciously, we decided to let a great technology with even greater limitations change our cities and the natural landscape to make it more suitable. While it opened many economic opportunities it also came with a high financial (building, fueling, repairing, insuring cars, paving, and maintaining roads, etc.), cultural, and environmental cost.

## CHAPTER'S KEY POINTS

- While AI is often associated with self-aware superintelligent machines and shiny humanoid robots, AI is primarily a statistical model using pattern recognition technologies.
- Thanks to Machine Learning techniques, AI is able to solve complex problems without being programmed with a specific set of rules.
- The power of AI is the outcome of the favorable convergence of a number of factors, such as high computing power, the availability of large datasets, and real-time connectivity.
- AI has considerable intrinsic limitations that should be kept in mind as its use continues to expand. For example, AI is unable

to understand basic principles such as cause-and-effect relationships and tends to do very poorly in any type of unpredictable environment.

- A serious assessment of AI's biggest strengths and weaknesses is needed to identify which activities are better suited for machines and which ones are better suited for humans.

# When Machines Appear to Think

Machine intelligence?

What levels of intelligence have machines reached? Will their intelligence ever surpass human intelligence? Throughout history we have often been found comparing ourselves against the latest technology. For example, when precision clocks were the newest machine around, some philosophers compared humans to mechanical and deterministic machines such as clocks. Now that we have much more sophisticated machines it is difficult not to compare artificial intelligence against human intelligence.

The use of the word "artificial" to describe machines' intelligence is straightforward: it is artificial because it is the outcome of an artifact, a man-made object. What has triggered the most heated debates is the concept of intelligence. The divergent views around the term intelligence are mostly due to the fact that it is an area in which we still know very little, and it even lacks commonly accepted definitions. While intelligence is undoubtedly one of the most exciting products of the universe, it is something that is still not clearly understood. Neuroscience has just started its quest to explore our cognitive abilities.

If we believe that the only form of intelligence is human intelligence, then it is obviously impossible, by definition, to have an artifact that can be called intelligent. If we are open to a more inclusive definition of intelligence, which could potentially include the cognitive abilities of other living organisms (on earth and beyond), or even machines, we need to then define clearly what the required components or qualities of intelligence should be. The components of intelligence most frequently referred to are self-awareness, ability to perceive and understand the world, to feel emotions, etc., but we do not have a set of commonly accepted standards and thresholds. What we can confidently say is that if intelligence is defined as the ability to understand the world, then AI has zero intelligence because machines cannot understand anything, not even basic concepts such as cause and effect. If intelligence is defined as just the ability to solve complex problems, then AI is already more intelligent than humans with regard to specific types of activities. If there was an arithmetic or image recognition competition between a smartphone and me, I would probably lose every single time.

Machines are typically described as intelligent when they have the ability to perform human-like cognitive tasks and the ability to solve a problem that we consider complex. It doesn't mean that machines have the same intelligence as the one that humans have or that tasks are executed the same way humans would, but the type of tasks are the ones for which we would expect some level of human intelligence. This is an important distinction because even if AI lacks the type of intelligence we are familiar with, it does not necessarily mean that it cannot perform tasks that we deem complex.

AI's intelligence is commonly grouped into three different levels. Narrow AI (or Weak AI) is a specialized model able to perform only one very specific set of tasks (arithmetic, speech recognition, image recognition, driving). The next level up is Artificial General Intelligence (AGI, Strong AI or Human-Level AI). AGI is the hypothetical scenario in which a model is able to match humans at all cognitive tasks, including the ability to solve any problem that we would consider complex. Finally, the highest level of intelligence is generally called Superintelligence. That's the hypothetical scenario in which a model is able to outperform human intelligence and outperform all humans on earth combined. Where are we today?

Narrow AI is already everywhere. For instance, this is the type of AI that runs web search engines, that monitors our credit card transactions, or that makes recommendations for our streaming services. In certain specific (narrow) areas, AI already outsmarts humans. For example, AI is already much better than humans in arithmetic, certain board games, or tagging faces on social media.

AGI does not exist yet and it is unclear if it ever will. Some researchers believe that it will happen within just a few generations. Some believe that we will only have AGI once we are able to create a system that is able to mimic the human brain. Others believe that the intelligence produced by our brain is just one example of intelligence and that it is quite likely we will get to AGI by discovering new types of intelligence that work differently from the one we know. It should be noted that given the astonishing advancements made with AI-enabled chatbots, many have described such applications as the first spark of AGI. It is yet to be seen if those models have in fact any features that are qualitatively different from any others seen so far.

With regard to Superintelligence, the general public tends to associate it with the idea of a machine that tries to take over the world and enslave all humans causing a permanent evolutionary stagnation or their extinction. Most researchers are skeptical about such a concept and consider it just a distraction from more imminent and certain risks. Others consider it a serious possibility especially if we fail to timely deploy controls to keep AI in check.

## Are we heading toward Artificial General Intelligence?

The most popular questions around machine intelligence are around the expected timing for the development of AGI. The researchers that met at the famous conference organized by the Future of Life Institute in Puerto Rico in 2015 agreed that AGI will happen by

2060, but the answers tend to greatly vary across experts, ranging between just a few years to millions of years or perhaps never at all. The topic about the timing of AGI, while popular, is probably one of the least interesting ones for a number of reasons.

First, due to the lack of a common definition of AGI, the debate around the timeline for AGI has been mostly meaningless as based on different interpretations of AGI. Focusing the debate on the definition of AGI itself would be a much more valuable effort, at least as a first step.

Second, we have already proven to be bad with predictions around technology's timelines because we tend to greatly underestimate and overestimate our own abilities. When the first group gathered in 1956 to explore AI, they genuinely thought they were going to need six to eight weeks to figure it out. Six decades later, we are still very far from it and still debating its definition. In other areas we have overestimated the timeline. When Nobel Prize winner, Lord Rutherford, commented on Sept 11, 1933, on the idea of generating energy using atoms he stated that "anyone who looks for a source of power in the transformation of the atoms is talking moonshine." About 16 hours later, on Sept 12, 1933, Leo Szilard demonstrated a way to do exactly that.

Third, human level intelligence has already been achieved in many areas, and in several of them it is already way beyond human level. Can anyone beat their calculator? Can radiologists beat the image recognition abilities of their algorithms? (They can't). A narrow version of superintelligence is clearly already here, and the number of areas in which machines are already better than humans continues to grow with more and more applications. Additionally, given how fast technology evolves, the step from human-level intelligence to superintelligence in those specific areas tends to be almost instantaneous.

The Narrow Superintelligence table is a simple visual created to help guide the discussion around machines' intelligence by focusing on the level of intelligence reached by machines for individual types of cognitive abilities rather than just on how far we are from AGI. The ten columns are just examples of the thousands of abilities we possess (arithmetic, image recognition, etc.) and the level of fluency that has been achieved by machines for those specific abilities.

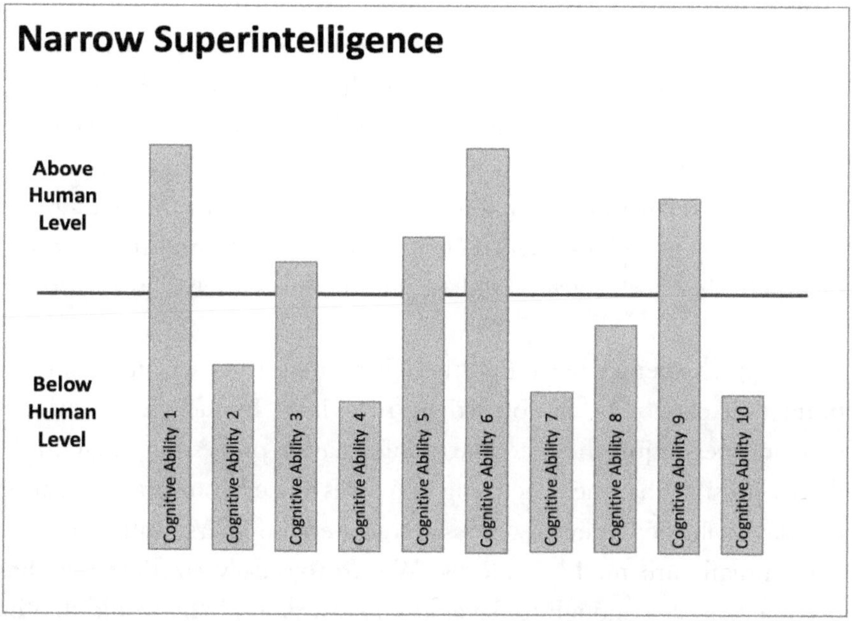


What really matters is the machines' evolution across all areas because once machines reach human-level capabilities in a significant number of them, the jump to AGI will happen incredibly quickly. Rather that focus on the AGI timeline, it is much more interesting and useful to assess what types of human activities are expected to be performed particularly poorly or particularly well by machines so that we can prioritize the areas where our attention is going to be needed most in terms of public policy, safety research, regulations, etc.

## Machines' consciousness?

Will machines ever achieve consciousness, with the ability to think, be aware, and have feelings such as love, happiness, fear, or pain? Will they ever develop a sense of personal values or even some level of moral judgment? In other words, will there ever be a moment in which a machine will say something like "I'm afraid, Dave", as seen in Stanley Kubrick's classic film 2001: A Space Odyssey, while the machine was being shut down?

That's a very difficult question because the nature of consciousness is very little understood, not only for machines but also for humans. We know that consciousness is the outcome of the activities that take place in our brain, but for the most part we still do not understand how they take place and we do not know if there are activities that cannot be replicated by machines. Neuroscientists are trying to find that out, but the quest has just started and at this point we have just a lot of hypotheses.

Some experts believe that there is nothing truly special about our brain that cannot be performed by machines. In such a scenario, if consciousness is just the outcome of information processing, then it is in theory possible that the processing that takes place in our brain might be one day replicated or even surpassed by supercomputers. Both machines and humans are made of atoms. We are basically stardust that has become complex enough to develop self-awareness. Why would that be different if the same complex patterns exist in machines? Why wouldn't similar computing activities generate similar results? All things considered, machines and humans have a lot of things in common: we have sensors, we have memory, we come up with models based on the data we capture, and we control everything with a central control unit. If we create a machine that is also able to fully replicate a human neural network, it is difficult to find a reason why it would not gain consciousness.

Other experts believe there might be unique biological properties of the brain that cannot be replicated by machines. In this view, our brain is able to provide features, including consciousness, that would never be matched by machines. Intuition and creativity, for example, are frequently mentioned as qualities that are uniquely human, that machines will never be able to display. There have been, however, many examples of machines creatively finding new solutions, such as the one of Google's AlphaGo playing against the world's champion of the board game GO. In another interesting case, a self-driving car independently identified a creative solution to signal to other cars to go ahead when stopping at a four-way stop sign intersection: without being programmed for that, the algorithm determined that backing up a few inches after stopping is an effective way to encourage other cars to go first. Another quality we tend to assume is uniquely human is intuition, generally defined as the ability to make the right decision despite a lack of evidence by "listening to our gut". The reality is that there is no evidence that intuition is any different from just luck. Guessing is re-labeled after the fact as intuition only for those cases in which the guessing was correct. Considering that machines are used to make decisions affecting humans' lives, do we even want machines to guess rather than use data?

What we know for sure is that, at least for now, we are very far from having machines that are able to replicate the complexity of our brain. Even if in the very far future we get to self-aware machines, it will not happen overnight. It is reasonable to assume that there will be a number of development stages where we will have the opportunity to test machines' abilities and progress over time.

The most famous test for measuring machines' ability to think like humans is the Turing Test or imitation game, developed by the English mathematician Alan Turing in 1950. In the test, a human is asked to submit written questions to two participants in different rooms, and to determine which one is human based on their written replies. The two participants' goal is to convince the tester that they're human. If the machine can convince the tester that it's human, then it has demonstrated the ability to think. The assumption, and the test's own limitation, is that the ability to think is considered achieved once a human misclassifies a computer as human. The problem with the test is that, even if passed, it does not necessarily prove the ability to think, but only the ability to mechanically match questions to the appropriate answers, or perhaps the algorithm's ability to simply deceive the tester without understanding anything about the questions and the answers.

Additionally, given the recent advancement in the technology behind chatbots, the text generated by AI is now so impressive that it can easily be flagged as non-human, not because of its weakness but because of its strength. For instance, if we asked for a list of all characters mentioned in the Bible in alphabetical order, a chatbot's response would be so instantaneous and complete, that it would be easy to tell it did not come from a human.

Even just choosing the right terms to describe the characteristics of AI can be particularly problematic because even small differences in the definitions and the criteria used in the assessment can lead to very different results, with the risk of creating confusion around what machines are able to do. For example, if machines process data in a way very that is different from the way our brain processes its thoughts but the result of such processing is qualitatively comparable to the product of human thinking, is it more correct to say that machines can think (in consideration of the quality of the results) or is it more correct to say that they cannot think (considering the differences in how they operate)? Rather than getting caught in such a theoretical discussion, we should instead focus on the type and quality of tasks machines can perform. If we think of an airplane, it would not make much sense to evaluate its qualities by comparing it to a bird. What matters is that the airplane can fly as intended even if it is very different from a bird in its physical features and the way it functions. The same thing is true with the discussion around a machine's ability to think. As Dutch scientist Edsger Dijkstra famously put it, "The question of whether a computer can think is no more interesting than the question of whether a submarine can swim."

Looking ahead, as AI gets better and better at simulating intelligence and consciousness, we are going to find a lot of situations where humans believe they are interacting with something self-aware when in reality it is just an AI model working to reach a programmed goal. As well described in the Ex Machina film, if machines show feelings, it is likely just a statistical program providing the most common answers pulled from millions of examples in available data, but it is tempting to interpret them as a clue of something more than just math. Even experts will be subject to such risks of manipulation and confusion about the nature of our interlocutors. For example, in 2022 a top researcher at Google interacting with a chatbot was led to believe that the algorithm had come to life just because the model came up with answers that made it sound human. Additionally, since we are creating a technology that resembles something alive, it is not unthinkable there will be concerns from certain groups about the ethical implications of having humans "playing God" by creating life-like beings. The most interesting part of the debate around that question from an ethical perspective, rather than whether it is OK or not to play God, will be whether something that looks (and says it is) alive should be considered and treated as alive. With the understanding, however, that no matter how close AI will ever get to human-like qualities, there will always be one thing that will separate us from machines: machines will never be bound to the same cycle of life (from birth to death) to which all humans are bound. Can something be considered alive if it was never born and not necessarily destined to die?

### Humans no longer in the loop and responsibility gaps

One of the most powerful features of ML is the ability to act autonomously with minimal or no direction from humans. While a machine is still executing a program, some of the actions taken are not fully predetermined by the machine's programmers as the machine has the ability to learn from its environment and use such information to maximize its goal.

There are different degrees of autonomy in which humans' real time oversight can diminish all the way to non-existent.

In some applications, human feedback is still needed along the way to provide direction, validate information, and make decisions. For example, AI might be used for robotic surgery under the constant oversight of medical professionals to confirm recommended actions, before they are performed by the machine.

In other applications humans are not needed to complete the task but they are kept constantly in the loop or at least on the loop with the ability to take over as needed. Most of the autonomous machines we are familiar with such as self-driving cars, or the ones used for space and deep ocean exploration, fall into this category. It should be noted that having humans in the loop does not necessarily result in meaningful engagement or active participation, especially when humans' approvals are somewhat repetitive and dependent on other types of machine-driven assessments. For example, if the approval of a certain activity becomes an ongoing repetitive task (such as approving the launch of a missile whenever recommended by an AI model monitoring air space), the person in the loop might end up automatically approving the recommended course of action without meaningful examination. There might also be a psychological aspect that acts as a discouragement for humans to take action. For instance, if the dosage of a certain medication is autonomously determined by an AI model, the doctor in charge might decide to approve the dosage just to avoid further scrutiny, paperwork, or just out of fear that a deviation from the AI recommendation could be seen negatively if the case ends up in a courtroom. There might also be situations in which there is no sufficient time for people to take action. When the self-driving SUV ran over and killed a person crossing the street in Arizona in 2018, the person inside the vehicle only had a second or two to take control of the vehicle, which was tragically insufficient to avoid the crash.

In some other applications machines operate for a considerable amount of time without having humans in the loop. Lethal autonomous weapons and self-driving cars are perfect examples in which AI has already reached a point in which they are fully autonomous without anyone in the loop supervising the machines. By not having humans in the loop in real time able to intervene as needed, fundamental ethical questions inevitably arise, especially when machines are put in situations to execute plans that might result in loss of life.

The more autonomy is given to machines, the more unclear it becomes where the moral and legal responsibility lies for any harm that they might cause. If the execution of the activities that cause harm are delegated to machines and the humans who deployed them are not aware of what is happening and have no control over the outcome of those activities, who should be considered responsible?

Given that machines lack consciousness or free will, we obviously cannot just blame them for the harm they cause. The responsibility lies with the humans that built and deployed them. However, this might not be as easy as it sounds. First, if humans are not fully in the loop it might be difficult to demonstrate their knowledge of the risk of their ability to control the outcome. Second, it can be very difficult to identify a responsible party considering the whole chain of human decisions behind the development of machines (such as all the different players that develop, build, test, and maintain the machines, their hardware, their software, and so on). Third, holding specific people accountable might be very difficult when harm is caused not only by just the way the algorithm works, but also by the type of the data it is fed with or perhaps technical characteristics around the quality of the machine's software or hardware, such as the reliability of its sensors. This inevitably creates a serious responsibility gap where no one feels directly responsible, and no one is held accountable.

Will humans become too dependent on AI and ultimately obsolete?

Important matters in our lives are already being delegated to machines. It is reasonable to assume that the trend will continue, creating some sort of "Robocracy" in which AI will be in charge of some of the most crucial aspects of our lives. This trend will come with a number of risks.

For example, as more tasks are delegated to machines, there is a risk of becoming too dependent on them for vitally important activities. Is our future like the one represented in the Pixar movie Wall-e, where humans are so well taken care of that they become completely reliant on machines and are not able to do anything themselves? Especially with autonomous vehicles, there will be an increased reliance on the machines themselves to a point that it might become safer to exclude human control altogether. The autonomous vehicle industry is now debating whether it is safer to keep humans in the loop or avoid any expectations for them to respond altogether. For example, while most manufacturers are developing cars that still allow humans to take over even if in auto pilot mode, some others believe that it is unreasonable to expect humans to successfully take over without proper warnings and therefore it is safer to completely exclude them from the car's controls, and to design its functionalities accordingly.

From a more philosophical perspective, as more human abilities are surpassed by machines, we might become increasingly obsolete when compared to the power of AI. With so many functions successfully delegated to machines, it is natural to start wondering if there is something special that humans have that will always put us in a position of superiority no matter what AI is going to be able to do. If so, what is it? If not, will humans at some point become less and less relevant and eventually become completely unnecessary?

The trend has fortunately been very slow, but we will need to avoid finding ourselves like a frog in a pot enjoying the warmth of the water without realizing it is being brought to a boil.

## CHAPTER'S KEY POINTS

- If we want to learn more about the potential of machine intelligence, we need to better understand human intelligence.
- If intelligence is defined as the ability to understand the world, then AI has zero intelligence because machines do not have this

ability. If intelligence is defined as the ability to solve complex problems, then AI is already more intelligent than humans with regard to specific types of applications.

- The type of intelligence achieved so far by machines is commonly described as Narrow AI (or weak AI) given its ability to be deployed for a very specific and narrow range of tasks, as opposed to the hypothetical concept of Artificial General Intelligence (AGI or human-level AI) and Superintelligence.
- There is no scientific evidence that machines will ever achieve some level of consciousness, but there is also no evidence that they will never get there.
- Despite the level of autonomy AI has reached, having humans in the loop in meaningful ways continues to be important especially in situations that affect human life.
- As more tasks are delegated to machines, there is a serious risk of becoming too dependent on them, especially for vitally important activities.

# Best of Times, Worst of Times

Tale of Two Cities

Is AI the best thing that has ever happened to us or the worst? The most likely scenario is that it will be both.

When I think about the AI revolution, there is a sentence that describes very well where I think we are. In 1859, the famous writer Charles Dickens wrote a novel called A Tale of Two Cities set in London and Paris before and during another revolution, interestingly, the French Revolution. The book is not particularly relevant, but the opening really describes well our world today: "It was the best of times, it was the worst of times, it was the age of wisdom, it was the age of foolishness, it was the epoch of belief, it was the epoch of incredulity, it was the season of light, it was the season of darkness, it was the spring of hope, it was the winter of despair."

On one hand it is impossible not to be excited about the future. AI might very well be the best thing that ever happened to humanity. Intelligence is our best tool to solve problems and having more of it might turn out to be instrumental in solving today's biggest issues such as diseases, malnutrition, and climate change. AI might empower humanity and help it flourish like never before. AI can help us move faster with science and new discoveries and give us more of what we value the most: time. As AI continues to evolve in the upcoming decades and centuries, more and more jobs will be performed by machines. This might mean that we will finally have more time to do what we really want to do: we can let the machines grow food and build houses, while we spend time with friends and family, reading, listening to music, etc. At some point in the far future, we might even change the way we see unemployment. Unemployment today is a sign of poverty. In the future it might be a sign of a high quality of living. If you have to work, it may mean you are in a technologically underdeveloped country. Looking at the far future, if we are going to start colonizing the universe within our galaxy and beyond, it is difficult to think it is going to be possible without some type of AI-driven colonization probes. The prospects are so promising that it would be unethical not to pursue them.

On the other hand, many have raised serious concerns about AI threats. Stephen Hopkins, for instance, described AI as the biggest existential threat to the human race. Elon Musk has described AI as potentially evil and even more dangerous than nuclear weapons. Bill Gates publicly stated while talking about AI, "I don't understand why some people are not concerned". Those are not folks typically against technological innovation and are generally considered some of the smart ones. So then why do they see such apocalyptic scenarios? Is it perhaps because they are so wealthy that they do not really have anything else to worry about? Or is it perhaps a clever way to distract the public from the issues that are happening right now such as the huge accumulation of data and lack of privacy, so that the big firms keep doing what they are doing undisturbed, making decisions behind closed doors?

Whether those concerns are founded or not, there are still a lot of unanswered questions with regard to AI. For instance, are we sure that AI will be beneficial to us individually and as a society? If AI will be on point for almost everything, how will we make sure AI will operate in line with our values? Are we sure the economic benefits will be fairly distributed among everyone? How are we going to handle the transition toward that state? What if this immense power gets in the wrong hands?

What if AI gets completely out of our control threatening our own existence? Are we sure we will always have a way to turn it off?

What we know is that AI will continue to revolutionize our lives. Many experts see AI as potentially more impactful than the invention of the printing press and any other technological revolution. The opportunities are unprecedented, but so are the ethical implications and risks, from questions related to privacy and transparency to social injustice. Let's examine some examples of sectors and functions already going through massive transformations thanks to AI.

## Health Care

Many of the most exciting improvements driven by robotics and AI are in the health care sector.

In the diagnostics field, doctors can take advantage of machine learning's high resolution image recognition capabilities to detect diseases early, leading to better diagnoses and helping avoid the need for invasive procedures such as biopsies. Additionally, AI software has already been proven to be more accurate than humans at recognizing patterns in retina scans to detect diabetic retinopathy or cancer patterns in x-ray images. Black boxes are so good at recognizing patterns from x-rays, with accuracy rates that are often higher than the ones of humans, that some experts have already proposed we reconsider the curriculum for aspiring radiologists. An additional advantage of AI-driven diagnostics is that it helps address doctors' biases. For instance, research shows that male doctors tend to misdiagnose heart attacks in women much more than they do for men, mainly due to preconceived notions about women's response to pain. With the right data points, AI can help eliminate such a bias.

The ability of AI to analyze large amounts of data can help discover new medications, and better assess medications' qualities, such as effectiveness, most efficient dosage, and expected side effects. Not only will AI help discover new medications such as new types of antibiotics, but will also take a second look at solutions that were dismissed due to unsatisfactory tests. For example, AI will help find correlations never seen before and help nail down the causes of adverse reactions in certain groups (example: uncommon genetic characteristics), opening the way to using those medications for all other groups. AI's ability to analyze large amounts of data will also be game changing in regard to DNA analysis, allowing doctors to find patterns across a vast population, helping us find new ways to predict, detect, and cure diseases. Given the vastness of genetic data AI is a perfect match to find correlations that can help improve research and medical diagnoses.

The vast amounts of data on individuals and the broader population can also be used to create risk scores to improve health management practices by leveraging predictive analytics beyond just traditionally used data points such as age and gender. Such risk scores can be used, for example, in hospitals for triaging patients visiting the emergency room.

Nurses and other medical professionals tasked with manual and labor-intensive work, will soon be able to receive help from robotics and AI as well. Wearable augmentative devices to enhance physical performance such as exoskeletons will help provide extra physical strength while completing physically challenging tasks such as moving patients to different hospital beds. This will also help reduce fatigue and on-the-job injuries. Autonomous robot-nurses will also help medical staff with low complexity repetitive tasks such as bringing supplies to hospital rooms or medications to patients. Such technologies, like the Aethon TUG hospital portering robot, have already made their appearance in several hospitals around the world. Robots will also play an increasingly relevant role with surgeries, either by assisting surgeons in the operating room or by performing surgeries themselves with minimal human oversight. As doctors' physical presence will no longer be necessary, this technology will likely help bring down physical barriers, to make the world's top surgeons accessible across the globe.

Robots will also be used more and more in therapies aimed to stimulate specific kinds of interactions to improve mental health, similarly to pet therapy and hippo-therapy. There are numerous examples of robots already used in pediatric wards to help kids that cannot have pets. The robots are able to sense the kids' feelings and help them relax, increasing the likelihood of successful recovery.

Care robots such as Willow PR2 facilitate the care of patients at home as well, helping people with bathing and feeding, providing reminders, and interacting for companionship purposes.

Another area where there has been a huge improvement is smart implants and biosensors. Some implants are nothing less than miniaturized chemical labs placed under the skin. They are able to analyze vitals, release medications, and communicate wirelessly to doctors for the exchange of data in real time. For example, there are machines today that analyze blood sugar levels and autonomously inject insulin in the bloodstream. The data collected across the patient population can also be shared across the medical community and used for research purposes. Considering that some of the data, such as heartbeat rates, can now be detected by wearables, it is safe to say that the amount of available data and related applications will continue to grow. Even data on the way people drive or type on a keyboard can be collected and analyzed for medical purposes allowing early detection of neural degenerative diseases. These types of technologies not only result in better care but also in a substantial reduction of time spent in the hospital. The monitoring of patients' physical and psychological states will allow doctors to track patients' conditions and deploy help to their homes as needed. The advantages seen with "from home" economy for learning, working, shopping, etc., we will therefore expand to the health care sector as well.

Within medical facilities, AI-enabled object recognition technology will help monitor compliance of safety protocols such as for instance tracking completion of personnel routine tasks such as applying hand sanitizer as they enter a room. AI monitoring will also allow doctors to have more precise insight in real time to patients' psychological well-being, allowing them to act in situations of stress that have no otherwise-apparent symptoms.

AI-enabled prosthetics will be particularly promising for improving the quality of life for individuals currently unable to move the upper and lower parts of the body and/or to effectively communicate due to disabilities. The next generation of prosthetics being explored is an implant connected to the brain to enhance our cognitive abilities, increase our memory, assist with mobility, etc. Tests on animals have already proven to be very successful allowing animals to complete tasks such as playing video games using their brain rather than manual controls. AI will ultimately allow us to decipher patients' thoughts by scanning their brain, providing new forms of communications for those who have lost the ability to speak.

Having a neural interface will also address one of the biggest bandwidth constraints in human-machine interactions: while machines can communicate at the terabyte level, human output is limited to just a few hundred bits per second due to the limitation of having to use basic tools such as keyboards operated via our fingers. The more our technology is integrated with our biological body the faster we will be able to interact with machines. If you look at human evolution, the newest area of our brain, called the neocortex, has given us the computational power that has allowed us to store more information about our environment and make better decisions. The growth of that area stopped once it became too challenging for childbirth, but in the far future AI may give us an opportunity to continue to expand it by using a synthetic cortex connected to our brain, giving us more computational power than we have ever had in a sort of hybrid between biological and synthetic information processing.

Another exciting application of AI and robotics in the medical field is the use of robots as avatars for people with impaired mobility such as tetraplegia. Thanks to AI, patients confined to a bed will be able to operate a robot with the use of just their eyes and experience the world through the robot's sensors, move around, interact with others, and even perform tasks. In Japan, for example, there is already a cafe staffed by robot waiters that are controlled remotely by paralyzed people.

The use of AI in the medical field comes with substantial risks and ethical questions as well.

With regard to the use of connected devices such as smart prosthetics or implants there is a serious risk of having patients' private information captured and used without consent. In addition to privacy risks, there might be safety issues related to the connectivity of the devices to wider networks, such as for instance the risk of having life-saving devices hacked. In the case of devices directly interfacing with the human brain to interpret its brain activity, there would also be the risk of having access to thoughts against the patient's will for purposes beyond the person's medical needs such as for law enforcement investigations or for campaigns against political opposition.

Another set of challenges are related to the interactions between patients and machines. For example, many fear that systematically delegating care to robots is not ethical as it deprives patients from the warmth of human interactions. Many are also uneasy about scenarios in which robots might have to use some form of psychological manipulation or physical coercion while performing their duties. For instance, do we always expect a robo-nurse to tell patients the truth about their medical conditions? If the robot is employed in a hospice facility, should it be direct with patients in regard to their life expectancy? Additionally, if a robo-nurse is on point for giving medications to uncooperative patients, there may be situations in which the robot might need to exercise a certain level of emotional manipulation or even physical force. Are there situations in which forms of physical restraint exercised by machines would be ethically acceptable? What if the robot was simply instructed to ensure a patient with severe disabilities or dementia timely takes lifesaving medications? And what if the patient being monitored by a robot runs away from a medical facility toward a dangerously busy street or tries to attempt suicide?

The lack of transparency in AI-driven black boxes used in the medical field has also led to a number of ethical questions. As the recommendations provided by the AI model might lack explainability, many have expressed concerns about making medical decisions based on determinations made by machines rather than humans, especially if the complexity of the model does not allow doctors to understand the reason for a certain recommendation made by AI. Fully delegating that function to AI also removes the opportunity for doctors to apply their own judgment to assess risk-reward considerations. For instance, if an algorithm recommends with high confidence, let's say 90%, a certain medication for a non-life-threatening condition, you still need a doctor to apply judgment to assess if a 10% error rate is sufficiently low considering the associated risk. If as a result of a misdiagnosis such medication would put the life of the patient at risk, the doctor might decide that a 10% chance of causing the death of a patient might be too high and opt for a safer, even if perhaps less effective medication. An additional element of risk is the fact that once a specific AI model becomes the standard in specific areas of diagnostics and is commonly used across doctors, it is going to be difficult to seek a meaningful second or third opinion from other doctors as most doctors will see the same type of diagnosis and recommendation.

There is also a risk that AI will end up providing lower levels of care to specific groups due to bias or gaps in training of the model or the way the model was designed. For example, a model trained to diagnose skin cancer might not have received sufficient cases across all skin tones, resulting in lower rates of success for people with, for example, darker skin, and may underserve or misdiagnose a whole group.

Finally, the better AI will become at supporting medical decisions, the more questions will be raised around accountability and professional liability. For instance, should the AI-driven x-ray software model turn out to be incorrect, will doctors using the technology still be accountable for the decision made using the machine's assessment? If not, should the machine's manufacturer or software developer be held accountable instead? Also, should a doctor decide to deviate from the recommendation made by a machine with a good accuracy rate, due to a doctor's personal gut feeling, should that somewhat increase the doctor's liability if the recommendation made by the machine was in the end proven to be correct?

### Human Resources Management

AI models are revolutionizing the way firms manage their workforce. For example, algorithms are now used for scanning resumes to assess who is eligible for a job interview. Forecasts on the candidate's potential success in a specific position are built comparing keywords included in the resume and historical data from the current workforce. AI can assist in monitoring the job interview, helping the interviewer spot non-verbal clues that are indicative of the qualities of the candidate. AI can even help managers identify who should be given a raise or a promotion, who is a high flight risk, who appears to be more creative or productive, or what teams have sufficient levels of diversity.

Using such tools to make decisions on such relevant aspects of people's lives comes, however, with many risks. For example, if they originated from AI's inscrutable black boxes, there might not be an adequate level of transparency or explainability on how such recommendations are developed.

Another risk is to have the AI model become a multiplier of existing company biases. For example, a few years back one of the largest companies using AI technology to scan resumes for entry-level applicants, quickly had to stop this approach when they realized that the model was discriminating against women. As the algorithm was using past performance data to guide resume screening and identify applicants with the highest likelihood of success, the model ended up discriminating against women by giving a negative connotation to certain female-specific keywords included in the resumes (such as "women's chess club") due to the fact that statistically speaking, women were less likely to get promoted or stay there longer than men. Using the available data, the AI model picked up that specific component from the company culture and started displaying the same unethical behavior.

## Education and Research

AI is already playing a substantial role in education across different areas. For example, the selection process for many universities now includes AI-driven models to assess applicants' likelihood of success in school and throughout their future careers. AI has also already started demonstrating its value in the classroom by creating new ways for students and teachers to communicate and collaborate. Additionally, AI can help automate many of the administrative tasks burdening teachers, allowing more time to interact with students.

Robots have already started being more and more present in the classroom. For example, robots are already being used by students that are unable to participate in classroom activities allowing a more natural interaction with classmates and teachers. The same opportunity, where the robot is a remote-controlled avatar, will surely expand to other sectors in which people are unable to have in-person experiences due to disabilities or other impediments.

Even scientific research will benefit from the superpower of AI by allowing researchers to shorten research cycles thanks to the ability to quickly analyze an enormous amount of data. For example, thanks to AI, a research team in Israel was able to instantly match fragments of ancient manuscripts that had been separated for thousands of years.

AI has also started benefitting a number of other fields such as biology and geology. For instance, DeepMind's AlphaFold was recently able to exponentially expedite the solving of protein structures, allowing biologists to make gigantic leaps in the field by mapping overnight what up until now would have taken an army of doctors decades to complete. With regard to geology, AI helps researchers not only to have better predictions of timing, location, and intensity of earthquakes and aftershocks, but also to discover new correlations between the geological properties that can trigger earthquakes.

The biggest risks seen so far in education are mostly around algorithm bias and lack of quality in the models. For example, models used to select college applications have already proven to be greatly prone to unfair practices due to bias in the historical data used to train AI. Some of the risks have already affected teachers as well. For example, algorithms currently being used in some school districts to assess teachers' performance have already proved to be grossly flawed, forcing schools to retire such models. Another emerging area of risk is around the use of AI to monitor students' attention and engagement via constant audio and video surveillance while in school or attending online classes from home.

## Financial Services and Insurance

AI offers a lot of opportunities in the financial industry as well. AI can, for example, significantly improve the customers' experience by offering customization options and real time interactions with virtual assistants. AI can also help firms optimize the underlying processes by automating a lot of the activities in support of customer needs. Above all, AI can greatly help with risk management activities such as fraud prevention by identifying outliers in customers' transactional data.

Relationship-based financial lending based on community ties and personal connections is quickly being replaced by models that look at a variety of data points. From items related to your financial health, total debt, repayment data, to much less straightforward details on applicants' backgrounds such as where they live, where they shop, or what they post on social media. For instance, a UK-based insurance company offered customers the opportunity to be considered for lower rates if they shared data from their Facebook account, as their algorithm allegedly managed to assess their reliability based on clues in their social media posts. For example, the more precisely people talked about upcoming in-person appointments (including space and time, rather than just saying see you tomorrow), the higher trustworthiness rate they received from the algorithm. In another well published example, Zest Finance created a model for assessing lending risk based on whether applicants used proper spelling and capitalization on the application form, how long it took to read the form, and if they actually read the terms and conditions (to assess how good they were as rule followers).

AI can also greatly help with early detection of fraudulent activities. By comparing multiple data points AI can help flag transactions carried out by unauthorized users. The model becomes particularly powerful when combined with biometrics, geolocation, and other data captured by personal devices such as smartphones. Even just the way we hold our smartphone during a transaction can potentially be used by a banking app to help identify situations in which the phone is not being held by the usual owner. Similar capabilities are also being deployed in the insurance industry. For example, with the use of tracking devices feeding location and driving habits in real time, firms no longer have to rely on average risk scores based on high level demographic data, and premiums can be customized based on a very accurate prediction of risk.

With so many opportunities also come a lot of risks. By having access to a large amount of data points about customers, it is not unlikely to have biased models that tend to discriminate against specific groups. For example, an algorithm deployed to assess the risk of a loan may end up using data points that are considered discriminatory such as an applicant's gender or ethnicity. As widely demonstrated, removing such data points from the model may not be sufficient to remove bias, as algorithms are very good at finding data proxies. For example, even after removing data related to ethnicity, the system might be able to use alternative data points, such as the applicant's zip code or first name, that lead to the same type of bias. Therefore, even if a model might not have any data on an applicant's ethnicity, it might still base its score on proxy items that represent a strong clue of an ethnic group.

Another instance of bias is for situations in which financial services are granted based on apparently non-relevant data points. For example, there are firms that assess the applicants' risk by measuring how quickly the application forms are filled out. This practice might negatively impact applicants who are not native speakers such as immigrants or perhaps applicants with disabilities, or parents, with little children underfoot. Given the large amount of data available we have already seen a few attempts by health insurance companies to predict individuals' future health care costs using risks scores based on a wide range of data points such as race, marital status, clothes size, time spent watching television, social media posts, or how timely bills are paid.

An additional issue is the opacity of black boxes used to make decisions. The complexity of the statistical model used, makes it at times impossible to explain how the data is processed and identify the determining factors of a decision. This lack of transparency is particularly troublesome for situations that fundamentally impact people's lives such as applying for a mortgage.

Finally, with so much data shared with firms, serious privacy concerns emerge across multiple areas of our life. Location trackers through car insurance plans are a good example of this type of risk where customers are offered discounts in exchange for sharing personal data. Will privacy become a luxury reserved for those who do not need to save on insurance premiums?

### Retail and Marketing

AI has already brought substantial advancement to the retail and marketing industry by increasing operational efficiencies and effectiveness. Grocery stores' checkout lines offer a very good example of how easily machines have started replacing humans. More and more stores have AI-monitored self-checkouts or are becoming grab and go stores with no checkouts at all, where customers simply walk out of the store. Where we still have humans, they mostly work as supervisors of multiple machines. AI is also widely used in stores to monitor aisles to automatically detect spills, empty shelves, and even shoplifting.

Even from a marketing perspective, algorithms can drive unprecedented accuracy in linking products to segment groups, to be then used in highly targeted advertising campaigns. Some stores already use smartphone apps to track shoppers' behaviors within the store to identify preferences and push customized ads.

The predictions are quickly reaching such accuracy that firms have deployed models in which products are shipped to consumers before they are ordered based on the data available about their daily life. For example, a sporting goods store might decide to ship a soccer ball to those who have recently purchased soccer shoes providing the option to return it for free. The error rate will soon become so low, it will represent a negligible cost of doing business.

The growing presence of personal digital assistants such as Amazon's Alexa, Amazon's home robot Astro, or sensors placed within refrigerators, will increase firms' abilities to become aware of consumer needs earlier than the consumers themselves. As the ordering (including price comparisons across different suppliers) will be mostly done by AI, even the marketing efforts will start focusing more on the machines in charge of ordering and less on humans.

Adopting AI models in those sectors comes with some serious risks as well. For example, the concentration of personal and meta-data in the hands of tech giants can easily result in violations of fundamental rights such as privacy and security. The same data might be used to create scores that end up affecting a certain group's ability to access goods and services. For example, unregulated technologies such as eScores are often used by online retailers to decide what items are available to specific consumer groups and to manipulate their prices based on a customer's profile.

## Agriculture

Agriculture is another field that will greatly benefit from AI. Thanks to more accurate weather forecasts it is going to be possible to maximize the use of scarce resources such as water, and to better prepare for inclement weather. For instance, the data collected by sensors around soil composition, temperature, and moisture, can help farmers accurately predict the best time to plant seeds, water, or harvest crops, increasing production and lowering the need for water and pesticides. AI will also drive efficiencies across operations by automating processes and reducing the need for human labor and even human supervision.

Among the potential outcomes for the transition to AI-driven agriculture, is the negative impact on workers currently employed to perform tasks that will be easily automated by machines, such as operating tractors or harvesting. Regions with high barriers to AI adoption such as lack of know-how or physical infrastructure will be penalized due to the lack of competitiveness against more technologically advanced global competitors.

## Architecture and Urban Planning

AI is helping architects make our buildings and cities safer and more efficient. For example, thanks to the Internet of Things (IoT) technologies with thousands of smart sensors connected to central systems, it is already possible to automatically adjust the lighting, heating, and cooling of buildings based on actual needs and environmental conditions, lowering cost and environmental impact.

AI will also revolutionize the way our cities operate, increasing the quality of life. For instance, AI-driven smart cities initiatives will help improve public transit thanks to the use of data to respond in real time to transportation needs and to communicate changes to passengers. Additionally, AI will help with disaster preparedness to predict effects of events such as hurricanes and floods. AI will also help increase public safety by allowing real time monitoring of public areas and automatically identify emergencies and prioritize responses. For example, in some US cities such as Cincinnati, AI is already used by Fire Departments to help triage emergency calls and optimize responses.

Automating such critical aspects of citizens' lives raises, however, a number of questions about the related risks. For instance, the collection of such a big amount of data through geolocation, facial recognition and similar technologies inevitably comes with important questions about privacy, data safety, transparency, and system auditability.

## Energy

AI will change the way energy is produced, distributed, and consumed. One of the ways energy productions will benefit from AI is via a combination of IoT technology and digital twins, where a physical machine such a wind turbine has a virtual copy (or twin) in a computer software developed to monitor the machine's behavior in real time via different types of sensors connected to a network. For instance, by analyzing the machine's noise or vibration, AI will be able to predict upcoming mechanical failures ahead of time and automatically order parts and deploy maintenance crews, dramatically reducing the down time needed for repairs. AI can also be used to maximize the turbines' efficiency by adapting the propellers orientation based on the orientation of other propellers and regional wind patterns.

AI will also help maximize distribution and consumption of energy by creating increasingly accurate predictions of peak times and by adjusting pricing in real time. Smart grid technologies, for example, allow bidirectional digital communication between energy suppliers and consumers which can be used to better respond to outages, to better integrate renewable energy, and to allow consumers to better manage their energy use and automatically take advantage of off-peak prices. This is particularly beneficial not only for powering homes but also for commercial facilities. For instance, the energy consumption of Google's power-intensive data centers is overseen by an AI model that helps reduce the energy load in real time. AI-driven robots are already used by energy companies in dangerous environments (such as the radioactive sections of power plants) or for the ongoing inspection of pipelines. AI can also help with the optimization of an energy production process. For example, Google's Deep Mind recently developed an algorithm that can be used to help control the plasma inside a tokamak of a nuclear fusion reactor. More importantly, the most promising research around clean energy production such as nuclear fusion, has greatly benefited from the use of AI, opening the door to powerful new ways to generate electricity.

Despite the numerous opportunities, introducing AI to the energy sector will come with risks as well. For example, the improvements needed for the transition to a greener AI-driven energy infrastructure will paradoxically put even more pressure on the Earth's ecosystem as it will require a higher output of rare minerals. Another risk has to do with security concerns. The more our energy infrastructure becomes dependent on AI, the higher the risk if bad actors hack entire systems causing disruptions of essential services.

## Environmental Sustainability

AI will deliver substantial benefits to a wide range of environmental efforts. Thanks to the large amounts of available data from satellites and ground sensors, AI-driven models can be instrumental in protecting our ecosystems from pollution, climate change, and loss of biodiversity. For example, thanks to satellite imagery, AI can be used for the early detection of sudden changes in vegetation due, for instance, to deforestation, or to monitor illegal fishing practices. AI will also help us create new low-carbon materials that can be used as an alternative to carbon intensive materials such as concrete and steel.

At the same time AI comes with high risks. AI will have a significant impact on our ecosystems due to the footprint of its software, hardware, its physical infrastructure, supply chains, energy production, water consumption, etc. For example, given the amount of data processed by machines, AI needs a massive amount of electricity. If such energy is not produced safely and sustainably, intelligent machines will soon end up being the biggest contributor to pollution. The cooling needed for data centers also requires a significant amount of water. Just as a point of reference, a single 15-megawatt data center running applications such as machine learning and cryptocurrency mining, needs up to 360,000 gallons of water every single day.

Additionally, the hardware required for machines to operate needs materials such as nickel, lithium, and cobalt, currently obtained from highly polluting mining operations often located in conflict zones. In addition to the environmental concerns, such operations are often linked to child labor practices and unsafe working conditions. So, while one side of the world might enjoy the best part of employing AI, the other side might be paying a high price from a human and environmental perspective. This paradoxically creates a vicious circle where the same technology that we are deploying to save the planet is among the biggest threats to the environment.

## AI-Assisted Government

AI can help with a variety of government functions.

First, it can help with policy development. AI could be used, for example, to better understand people's needs and identify the best public policy to address them. If our civilization is the outcome of our intelligence, having more intelligence in the decision-making process will likely work in our favor as it will help us make more rational decisions and foresee their outcomes, and even identify future needs. AI might end up projecting a greater authority than human expertise, and more easily gain support for the right public policies.

Second, AI could help governments with specific functions. For example, AI was recently used in a public health effort to prevent childhood lead poisoning. AI models were used to identify homes with a high likelihood of having lead paint to help prioritize the work of home inspectors. Other areas where AI, if used ethically, can be of great help are in support of law enforcement and the criminal justice systems. For example, the United States government now uses AI to flag tax documents with a high likelihood of tax fraud in order to prioritize audits. The biggest promise of AI in the criminal justice system is the reduction of subjectivity and bias. AI models are, for example, already used in judicial activities known for their high levels of subjectivity such as parole sentencing to assess in real time the risk of recidivism.

The risks associated with the use of AI by governments are particularly impactful as they are directly related to people's fundamental rights. For example, despite the promise of objectivity, AI can lead to biased outcomes that result in unfair treatments of specific groups. Additionally, the use of AI in critical functions such as sentencing, inevitably raises the question on whether it is ethical to have such fundamental decisions about people's lives delegated to a machine rather than a person. If a defendant prefers to be judged exclusively by people, should they have the right to choose? Also, given the opacity of black boxes, how can we comply with due process principles if we have no idea what process leads to a specific sentence?

## CHAPTER'S KEY POINTS

- AI might very well be the best thing that has ever happened to us, but it could easily become the worst if the many risks that come with it are not properly addressed.
- Every function and sector can greatly benefit from AI, from health care to education, but we have already seen countless situations in which AI can cause serious harm to individuals and society.

# Part Two
# **AI Ethics**

# Why Ethics?

Why do AI Ethics matter?

AI has become part of our everyday life. Whether we are aware of it or not, we use AI every day. AI-driven technologies heavily influence what we see when we run a web search, what we see on social media, what prices we see online, which news sources we read, what movie or song we stream, or what route we take when driving from point A to point B. AI's presence is so pervasive that it affects some of the most important matters in our lives. Whether we get accepted into college or get a job interview, or a promotion, or are approved for a loan, or considered at risk for certain medical conditions are just examples of situations in which our lives are affected by AI. The list continues to grow every day. For example, AI is now used to determine which neighborhoods police cars should patrol at night, and who should be released from prison on parole.

On one hand, the benefits delivered by intelligent machines are so promising that it would be unethical to turn them down. From safer transportation and better medical diagnostics to robots performing dangerous jobs, AI gives us the opportunity to build a much better world.

On the other hand, as we will see later in the book, AI comes with many different risks such as discrimination, invasion of privacy, social manipulation, lack of transparency, safety issues, value misalignment, labor misplacement, and even existential risk.

AI has proven so far to be an incredible superpower able to multiply the best and the worst aspects of our society. If we look at society's biggest challenges, such as inequality, climate change, political oppression, AI can easily be either a formidable solution or a powerful worsening factor. As a result, any effort to improve our world cannot ignore the power of AI as either the best ally or the worst obstacle.

To make it even more difficult, due to its versatility, AI is particularly exposed to the risk of being used for purposes that are different from the ones for which they were developed. For example, technologies developed for good reasons such as defense efforts, might be quickly converted into tools used to restrict civil rights. Similarly, technologies used for warfare scenarios can find very useful applications for all citizens (internet, GPS, etc.).

How can we ensure that all of this is done ethically and in line with our values? What is the right framework to ensure AI is socially responsible and truly beneficial?

## Ethical Frameworks

Ethics is the framework through which individuals and societies decide who they want to be, what goals they consider worth pursuing, and how they should achieve them.

At the individual level ethics guides our everyday decisions about our identity, priorities, and behavior. These choices are grounded in moral principles that shape our understanding of right and wrong and are closely tied to shared values such as life, respect, fairness, justice, honesty, and beauty. Values are highly influenced by our community's culture and often tied to social norms that historically have helped us live together, minimize conflict, promote stability, progress, etc. Some of those values translate into the moral obligations or duties we feel toward others (humans, animals, objects, etc.), starting with individuals in positions of authority or vulnerable ones such as children, the elderly, etc. Choices about how to treat others, what to do for a living, how to balance personal success with care for family members, are just some examples of decisions shaped by values that are concrete expressions of personal ethics.

At the community level, ethics helps define collective priorities that often become embedded in informal norms and formal rules. For example, when a town decides how to allocate limited resources such as funding for schools, healthcare, housing, or public safety it is making ethical judgments about its identity, what matters most and whom it seeks to protect or support.

Within a corporation, ethics defines organizational identity and long-term direction. Choices about which markets to enter, how products are designed, how employees are treated, how data is collected and used, and how profits are generated all reflect underlying values. In the corporate world, the term ethics is often avoided because it is perceived as abstract or secondary. Companies tend to prefer the word "strategy", which sounds more concrete and actionable. Yet ethics and strategy are inseparable and often different aspects of the same thing. Ethics defines the firm's identity, its long-term direction and its key principles, while strategy translates those ethical aspirations into concrete decisions and actions. In this sense, ethics is strategy viewed over a longer time horizon, and strategy is a way to operationalize ethical choices.

Ethics has three main fields of study: Metaethics, Normative Ethics, and Applied Ethics.

Metaethics is the study of the nature of ethics itself, how morality works in general and the concepts that are typically used to identify something as right or wrong.

Normative Ethics attempts to define standards that can help assess if something is right or wrong and what should be done in a specific situation. There are endless types of frameworks, but the three most common ones are Consequentialism, Deontology, and Virtue Ethics.

Consequentialism, which includes popular theories such as utilitarianism, focuses on the consequences of our actions and deems actions as ethical when they result in the greatest good for (or the least harm to) the largest number of people (even if harming others). Oversimplifying a bit, Consequentialism is all about maximizing happiness or minimizing suffering: it does not matter how we get there as long as the result is positive (all is well that ends well). Deontology, well represented by Immanuel Kant's theories, looks at our actions as a representation of a moral maxim and deems actions as ethical when in line with what we would want to be a universal law. What matters is not the consequence of the action, but how well the action aligns to what we deem right. The most common example of this maxim is "treat others as you would like to be treated". Virtue Ethics, often described as Aristotle's principle, promotes behaviors that demonstrate a strong character based on traditional values such as wisdom, justice, prudence, etc., rather than focusing on the consequences of the actions or on the sense of duty. The right behavior is therefore the one that a virtuous person would choose in the same circumstances and cultural context. Applied Ethics focuses on the arguments around specific moral questions and dilemmas. AI Ethics is the branch of Applied Ethics that focuses on the design and deployment of ethically sound AI systems.

The topics AI Ethics seeks to address can be grouped into two big buckets.

The first bucket is around what we want AI to do. How can AI help us with our most urgent priorities? What type of activities should be delegated to AI? Considering that AI can perform several types of tasks more effectively and efficiently than humans, this is ultimately about the opportunity to get the best out of AI by delegating to it anything that we would rather not do ourselves because we find it boring, repetitive, degrading, or dangerous, and anything that we want AI to perform, in order to better achieve our most important goals. Given the enormous opportunities we have in many sectors, from transportation to medical research, it is therefore not just about identifying the best use of AI but also about avoiding situations in which AI is underutilized, missing the opportunity to greatly help the human condition.

The second bucket is around what we do not want AI to do. Are there situations in which AI could be used in the wrong way or for the wrong reason? Some scenarios might involve situations in which AI causes harm because it is intentionally developed or used for illegal or unethical purposes such as government surveillance against political opposition or carrying out ethnic cleansing. Some other scenarios might include situations in which the harm is an unintended consequence of a legitimate use of AI, such as situations in which AI is not sufficiently safe, or ends up discriminating against a certain group, or contributes to poverty and socioeconomic instability. The same bucket also includes the ethical implications of delegating specific tasks to autonomous machines, in consideration of the fact that it might cause us to lose our own autonomy by losing control over the way they operate or by becoming too dependent on them.

## Moral Principles

The pursuit of responsible AI is typically tied to a number of key high-level principles often included in ethical guidelines by governments, industry associations and individual firms. There are hundreds of ethical guidelines touching on responsible AI all with similar key themes such as fairness, transparency, privacy, safety, etc. Research recently completed at Harvard grouped principles into 36 buckets, across eight key themes: Privacy, Accountability, Safety and Security, Transparency and Explainability, Fairness and Non-discrimination, Human Control of Technology, Professional Responsibility, and Promotion of Human Values. The main advantage of having high level overarching principles is that the core values can be easily applied across different technologies. This is particularly important in a field that is seeing such high rates of innovation, as it would not be feasible to reassess ethical requirements each time a new technology is discovered.

## Connect points with the Law

Ethical principles and laws are greatly interrelated with a wide range of situations in which they intersect. Ethical principles are not legally enforceable as such. In order to become enforceable, they need to be translated into actual laws so that they apply to all subjects regardless of their personal beliefs.

There are still several instances in which ethical views and legal requirements do not coincide. On one hand, there are situations in which legal behaviors are viewed as unethical. For instance, taking advantage of a tax loophole to avoid tax payments is legal but often viewed as unethical. Similarly, a law introducing racial discrimination will also be seen as unethical as violating fundamental values such as the principle of non-discrimination. On the other hand, there are ethical behaviors that are illegal. For instance, a mother stealing food from a store to save the life of a starving kid, while perhaps illegal, would be seen not only as ethical but even as a moral imperative. Lastly, there are situations in which individuals consider a legal standard too low and choose to do more than what the law requires. For instance, firms might decide to pay their employees more than just the legally required minimum wage to ensure they are able to support their families. This is important because compliance with the law is just a starting point. If I play a chess game, I am not going to win the game by solely following the rules of the game as I also need to play strategically.

All of this is also applicable to the world of AI. For example, when countries require by law the removal of certain content from the internet as part of their censorship programs, technology companies who consider censorship unethical will need to decide between the options of violating the law, leaving the country altogether, or complying with the law, based on perhaps other equally important ethical values such as the commitment to respecting countries' laws even when not in line with the firm's own values. Even just the way the law itself will be interpreted by a firm will have a high ethical relevance. These are true ethical dilemmas that can be addressed if firms have a good understanding of their values and the hierarchy of those values.

In terms of legislation, we are not starting from zero as some principles are already receiving some legal coverage. For example, principles such as privacy and AI safety are already partially protected through data protection laws and product liability requirements. However, as the capabilities and applications of AI continue to expand, there are vast areas still left untouched by legal requirements. For example, there is currently no law requiring independent assessments of AI deployed to execute sensitive tasks such as reviewing college and job applications.

Some firms have already started incorporating such principles into their internal policies. We still need, however, legally binding rules too as we cannot just rely on the goodwill of the players involved. Having public policy, laws, and regulations addressing our community needs and values not only can have a huge impact on safety and the overall trustworthiness of AI, but can also help promote innovation by creating clear rules for all players.

## CHAPTER'S KEY POINTS

- AI tends to be a multiplier of everything it touches with the potential of being both the best and the worst thing that has ever happened to us.
- Given the many risks and opportunities that come with AI it is imperative to develop the right ethical frameworks to ensure AI is socially responsible and truly beneficial.
- AI Ethics is the branch of Applied Ethics that focuses on the design and deployment of ethically sound AI systems.
- The topics AI Ethics seeks to address ranges from high level questions on what types of AI we should have, to more specific questions on what AI should be used for, related risks, benefits, and mitigation strategies.

- Ethical considerations should focus not only on the risk of misusing AI but also on the risk of underutilizing AI.
- Some firms have already started incorporating ethical principles into their internal policies, but they will need to translate into actual laws as we cannot just rely on the goodwill of the players involved.

# Misconceptions around AI Ethics

## Misconception: AI is Intelligent

When John McCarthy chose the word intelligence to describe this new class of computational systems while seeking funding for his 1956 workshop, the choice proved to be a brilliant marketing move. Benefiting from a name that suggested qualities deeply associated with the human mind such as understanding and reasoning, the term immediately captured attention, elevated expectations, and framed the technology as something far more interesting than anyone would have anticipated.

There is no question that AI systems can perform tasks that, until recently, required human intelligence. They translate languages, summarize complex documents, detect anomalies in medical images, generate convincing text and images, and solve sophisticated optimization problems at scale. However, the fact that they can complete complex tasks does not mean that they possess intelligence. It only means that intelligence is no longer necessary to complete those tasks. Even the most advanced systems often described as "expert-level" or "as smart as a PhD" lack attributes typically associated with intelligence such as the ability to understand, think, reason, etc. AI operates by processing data through mathematical models, identifying patterns, and optimizing outputs. It manipulates data without any understanding of their meaning. It generates outputs because similar patterns appeared in its training data, not because it understands what those outputs refer to in the world.

Imagine if the inventors of the combustion engine had called it artificial horses. The name would have been odd, but ultimately harmless, because it would have been immediately obvious that no horses were hiding under the hood. Even today, we measure engine power in horsepower without confusing that unit with the presence of animals. With AI, we are doing something similar, but the presence of the word "intelligence" is far more misleading. Intelligence is an abstract concept, and it is easy to assume that if a tool can complete complex tasks, there must be intelligence somewhere inside it. However, searching for intelligence inside AI systems is like opening a car's hood expecting to find horses.

This confusion is not incidental and it has clear beneficiaries. Presenting AI as intelligent simplifies marketing narratives, accelerates adoption, and supports high valuations. Recognizing AI for what it actually is powerful, scalable, but fundamentally mindless information processing does not diminish its significance. On the contrary, conceptual clarity is a prerequisite for responsible deployment.

### Misconception: AI makes decisions

One enormous misconception is around AI's ability to make decisions. Given AI's ability to operate within a certain level of autonomy, and sometimes even generate outcomes not fully foreseeable, many believe it can make decisions. However, AI does not possess what is needed to make decisions such as the values, independent judgment, or moral agency. While the use of AI and the way it works undoubtedly have strong ethical implications, it cannot come up with its own views on what is right and what is wrong. AI is nothing more than math used in sophisticated statistical models: it does not have its own agenda, objectives, ambitions, desires, dreams, ideas.

The impact of AI, with all its moral implications, is not about machines' ethical choices, but about the ethically relevant choices made by the people designing or overseeing those machines. What AI can definitely do, if instructed to do so, is to make a prediction about what humans consider to be ethical, but that's very different from having a machine developing its morals. AI's ability to solve complex problems cannot be confused with some type of motivation toward a goal. As pointed out by Steven Pinker, "being smart is not the same as wanting something". Practically speaking, there is no need to fear that your Roomba will someday elope with your smart thermostat.

When we colloquially say that AI "makes a decision" or perhaps "faces a moral dilemma" it sounds like the machine has some type of discretion. There is, however, zero discretion as the machine simply makes a data calculation that produces a score. There is an input (data), there is a statistical model, then an output (prediction). The distinction is important because attributing some type of moral agency to AI is not only incorrect but also makes it more difficult to create a culture where individuals and corporate entities are held accountable for the harm caused by the machines they design and to which they choose to delegate the execution of specific tasks. If AI causes harm, it is very tempting to just blame the machine, but there should be no one to blame other than the people who thought it was a good idea to design it and use it. Blaming the machine, while apparently harmless, tends to create dangerous confusion among the public on the source of the harm.

When we talk about what AI does, we are really just talking about what humans do with AI. Even when systems are designed in a way that is not fully deterministic, such as in the case of neural networks, this is really about the ethically relevant choices made by the people involved. If a movie streaming app has a recommendation algorithm that recommends an adult-only movie to a kid who has just watched a Peppa Pig cartoon show, it is because someone made a mistake and failed to consider certain risks, not because the AI model suddenly became unethical. If an autonomous car drives off a cliff it is not because it is evil but because someone did not take the time to properly test the technology before putting it on the road and perhaps because the public failed to request the right level of oversight.

There is no doubt that there are situations in which it is impossible for humans to fully assess the ethical implications of all possible outcomes of tasks delegated to AI, especially when based on opaque models, such as the ones based on deep neural networks. But even then, if the model is created so that there is room for unethical decisions you cannot blame the machine. If you take your dog for a walk, and you know its behavior is not fully predictable, the longer the leash is, the more risk you introduce for those you cross paths with. If the leash is too long to control the dog, or removed altogether, and the dog runs toward a kid and bites him, it is still your fault even if you did not instruct the dog to bite. This misconception is caused by multiple factors.

First, it is the outcome of decades of SciFi literature and movies with storylines where machines display human-like emotions and some type of moral agency. This seems to fuel our natural inclination to anthropomorphize everything that looks alive even when we know it is not, leading us to describe what machines do the same way we would with humans.

Second, mainstream news has also greatly contributed to creating confusion. For example, it is not uncommon to come across headlines describing algorithms using abusive language, lying, or even being racist, disregarding altogether the set of human decisions that led to such a design. Even the business decisions around the automation of processes, such as the deployment of AI and robotics to perform tasks in an office or a factory, are often misrepresented as something initiated by the machines themselves on a mission to steal jobs rather than something implemented by the humans running those operations. Other times the confusion created by the news is just an amplification of overly confident marketing campaigns around the capabilities of a specific tool aimed at attracting investments and publicity. By just regurgitating companies' PR statements, the news stories tend to overlook the tools' limitations and exaggerate their capabilities describing AI as a moral agent able to make decisions. The confusion becomes even greater when those news stories are accompanied by images showing shiny robots with anthropomorphic forms (like the robot on the cover of this book). If you wanted to graphically represent AI, it would be much better to show the image of an incomprehensible spreadsheet.

Third, such distorted perceptions are also due to the lack of proper terms in our day-to-day language to describe actions performed by machines, so that when we want to describe how a machine operates, we tend to use the same words we would use to describe human qualities or behaviors. Even words such as artificial intelligence, machine learning, and computer visions all have an evident connection to the human world. Similarly, it is not uncommon to say, for example, that a chatbot "hallucinates" when making a mistake or that a machine "believes" or "decides" something despite the fact that we know the AI model is just executing a program. While we would never say that a microscope is studying a microbe, we do tend to say that AI is driving a car and making decisions on how to best do it. It generally takes a long time for new vocabulary to form to describe a new phenomenon, so it's not surprising to see terms borrowed from different contexts. Other new disciplines face the same challenge. For example, neuroscience tends to borrow terms used in computer science, such as computing power and bandwidth, when describing the properties of the human brain. What matters is that when we describe AI with anthropomorphic terms, it is clear for everyone that it should not be interpreted literally. Saying that the sun rises or sets is fine, as long as we know that it only appears to do so as we rotate around it. Similarly, it is fine if we say that AI "decides" something as long as it is clear that it actually doesn't, as it does not have the ability to do so.

Even the way we describe the discipline of AI ethics itself pays the price for not having adequate terminology. For example, expressions such as AI ethics are often used interchangeably with other terms such as ethical AI, responsible AI, or trustworthy AI. The use of those terms tends to contribute to the confusion. With concepts such as ethical agriculture or ethical mining, it is very clear that the focus is on the ethics of the people involved. When we say ethical AI or responsible AI, however, it might not be clear that the focus is not on the ethics of the AI system itself but on the ethics of the people that design, deploy, and oversee the technology. The same way we do not have ethical calculators, thermostats, or toasters, we cannot have ethical AI. The term trustworthy AI is also less than ideal as it implies some type of mutual relationship between machines and humans. The trust should be in the people that develop and deploy the technology, not in the technology itself. For example, if we are confident about the safety of an elevator, it is not because we put trust in the elevator itself, but because we put trust in the people who designed it, built it, installed it, and maintain it, and the different rules and independent institutions created to ensure elevators are safe to use. If I get electrocuted while using a toaster due to a design flaw, the trust issue I am going to have is not with the toaster itself but with those who built it and sold it to me. What needs to be trustworthy is the whole ecosystem around AI, from those who collect the data, to those who design the models, test them, market them, deploy them, maintain them, etc. If I need surgery and I am uneasy about the operation, it is not because I do not trust the scalpel, but because I am not sure if I can trust the person performing the operation or the process as a whole. Saying that we need to make the AI more trustworthy makes sense as long as what we really mean by that is that we need to hold accountable all the people that play a role around the creation and deployment of that technology so that end-users can reasonably put their trust in the people behind it.

### Misconception: AI is ethically neutral

People tend to see tools as ethically neutral. The general view is that all technologies can be good or bad depending on how they are used: whether you have a hammer or a gun they are ethically neutral until someone picks them up with a specific goal. For example, AI-driven facial recognition technologies will be perceived as good when used to stop the bad guys and bad when used to stop the good guys. This is correct only if you limit your analysis to just the impact of a specific use. The reality is that such a view fails to consider all the ethically relevant decisions that precede the time a tool is finally used, such as the design and development phase. A tool cannot be considered neutral if it is the outcome of ethically charged decisions. While it is true that the purpose for which a tool is used is ethically relevant, it is also true that the ethically relevant choices made about the tool since its first conception are also embedded in the tool itself making it ethically charged even before it is used by someone. For example, you can drive a car to do something right or wrong, but the purpose you use it for is not the only aspect that is relevant from an ethical perspective, as you also need to consider its design, how it was built, its safety, environmental impact, socioeconomic effects on culture, etc. As another example, there are small land mines that resemble the shape of a green bird or butterfly that have historically been particularly effective at killing small kids because children tend to pick them up thinking that they are in fact a toy ("The Green Parrot" and "Butterfly Mine"). Because of this design choice, they have been considered even more immoral than other land mines. Once again, a technology can be morally charged just based on the way it works. There are also situations in which ethics play a role even regardless of how a technology works or even if it will ever be used. For instance, if you are against torture because you consider it a violation of an inalienable human right, any machine designed exclusively for torture will be unethical no matter how it works or where it is used. The fact that they exist as an option is something already ethically relevant.

The same thing is true with AI. Given the fact that algorithms are statistical models, people tend to consider them ethically neutral and inherently objective until they are used for a particular purpose. If you look, however, at how they are designed and how they work, there are plenty of decisions that make AI everything but neutral. While it is true machines objectively execute their calculations using the data they are fed and follow the rules they are given, AI's objectivity depends on the objectivity of such data and how it is processed. If a model uses racially biased data, the outcome will be biased as well. For example, if the data is sourced in a context in which minorities are discriminated against, the outcome of the model will inevitably reflect the same bias and even amplify such patterns. The more we believe AI is morally neutral the more difficult it is to ensure that people feel ethically responsible for the quality and safety of technology they create and their impact on society.

## CHAPTER'S KEY POINTS

- AI systems can perform tasks that, until recently, required human intelligence. However, it does not mean that they possess intelligence. It only means that intelligence is no longer necessary to complete those tasks.
- AI cannot make decisions, but only follow instructions as machines do not have their own agenda, objectives, ambitions, desires, dreams, or ideas. The impact of AI, with all its moral implications, is not about machines' ethical choices, but about the ethically relevant choices made by the people designing or overseeing those machines.
- Some of the confusion about AI's abilities are due to the lack of proper terms in our day-to-day language to describe actions performed by machines.
- AI is not ethically neutral as it reflects the ethically relevant choices made throughout its design, development, and implementation.

# Why is it so difficult?

## Multidisciplinary nature of AI Ethics

Why is it so difficult to align AI to our values? One reason is the complexity and the multidisciplinary nature of the AI ethics as a discipline itself. Addressing the breadth of everything AI touches requires expertise and insight across a wide range of disciplines such as computer science, data science, philosophy, law, statistics, psychology, ethics, sociology, risk management, international relations, public policy, economics, business management, and everything in between.

Understanding the breadth and depth of AI ethics is something that is difficult for anyone. For example, it is not reasonable to expect that software developers working on those technologies have the expertise to fully understand and address socioeconomic implications. Similarly, it is unlikely to find sociologists or attorneys who also have a good understanding of the AI's underlying technologies such as deep learning and neural networks. As a result, it has been difficult to bridge the gap between all relevant disciplines and create a methodical and comprehensive categorization of the different aspects of the ethics of AI.

## Nature of human morals

Another big challenge to the ethical deployment of AI has to do with the highly subjective and ever evolving nature of human morals. Our morals are the outcome of a combination of multiple factors such as personal experiences, the values embedded in our culture, and even our genetic makeup. The challenge of aligning AI to our values is challenging at multiple levels.

First, because as values vary across individuals, communities, and countries, there is not a one-size-fits-all set of values. For example, if I think about the way a self-driving vehicle should respond to accidents, and whose life should be prioritized (passengers, pedestrians, etc.), due to my personal religious or political reasons, my view might be quite different from another person's view, including members of my own family. Similarly, a self-driving vehicle programmed to be in line with the ethical views of a specific country might be considered unethical in other ones. For instance, I think that running over an animal is always preferred to running over a person, but that might not be necessarily true in certain cultures in which animals are considered sacred. Global companies will therefore need to assess what type of minimum ethical standards to elect across the globe and how to best integrate them with incremental standards that might exist at a local level. For example, a self-driving car designed to follow certain ethical behavior in one country might need to be re-set to operate in another country to align to local values.

Second, because values evolve over time, any type of framework in place to ensure AI is in line with our values will need to evolve over time as well. So, the assessment of the technology from an ethical point of view will need to be reassessed across the technology's lifecycle. The relationship between technology and ethics also tends to be bidirectional: not only our view of a specific technology will change based on the evolution of our values, but also our values themselves will evolve based on technological advancements. As technology brings new options, it changes societies and their ethical norms. For example, once synthetic meat becomes as accessible and affordable as traditional meat, it might become unethical to eat meat procured by slaughtering live animals because of cruelty, the environmental impact, etc. A similar type of evolution in our ethical views will take place as new forms of AI arise. This will not happen instantaneously as values require multiple generations to form and become part of our cultural makeup. Even for many ethical questions we face today, we have not had a chance to truly mature our view on what is right and wrong. With technological advancement accelerating even more, it will be very hard to keep up and have the right set of values to guide us. Additionally, our brain is not really set up to process data at the speed we experience today, so it is difficult to assess ethical implications of different options if there is no sufficient time to digest the information. Our brain makes better ethical decisions when responding to a letter than to an instant message.

Finally, we are not consistent in how we interpret and practice our own individual values. For example, our ethical views are easily conditioned by the specific context or conflicting principles. For instance, while most people consider it ethical for a self-driving car facing a sudden failure to protect innocent bystanders rather than the vehicle's passengers, most people also would refuse to ride in such a car given a choice. Our values are also conditioned by emotional or esthetic considerations. As very well stated by German philosopher Friedrich Nietzsche explaining how morals are influenced by esthetic criteria, "If you crush a cockroach, you're a hero; if you crush a beautiful butterfly, you're a villain".

### Complexity of programming ethics into machines

Another big challenge is around the methodology for programming ethics into machines to reduce the likelihood of unethical outcomes.

Many have tried to come up with universal ethical principles to be programmed into the machines. The most famous example is the attempt by science fiction author Isaac Asimov, who developed the so-called Three Laws of Robotics. The first law states that a robot may not injure a human being or, through inaction, allow a human being to come to harm. The second law states that a robot must obey the orders given to it by human beings except where such orders would conflict with the first law. The third law states that a robot must protect its own existence as long as such protection does not conflict with the First or Second Law. The three laws have been widely criticized due to their inability to cover a wide range of scenarios. For example, if a person decided to do something particularly risky such as horseback riding, a robot following those rules would be compelled to stop that person even if the person ordered the robot to step aside. The rule also implies the machine's ability to successfully predict people's behaviors, the consequences of the actions taken and potential alternatives, which is still very far from the current state of technology. Despite its gaps, Asimov's laws have helped promote a much-needed debate around AI ethics and how to ensure machines are aligned to our values.

Some experts believe the best course of action is to pre-program human values into the model itself so that it is given all instructions and guardrails needed to avoid unethical outcomes. For example, the algorithm assessing movie recommendations for streaming services can be proactively instructed not to ever recommend an adult movie when the viewer signs in with a kids account.

Others believe that providing instructions is not sufficient as the number of scenarios in which something could go wrong is just too high to be addressed ahead of time. According to such a position, it is more efficient and effective to let the machines observe our behaviors and identify our values by forecasting what a human would consider ethical. In addition to being technically very difficult, such an approach comes with the assumption that human actions accurately represent what ethical behaviors look like while we all know well that humans are quite inconsistent with regard to their choices.

## Lack of involvement of the general public

The general public has mostly been excluded from the discussions on the ethical challenges related to AI. The topic is still considered just a technical matter, and the debate is primarily led by large companies. Due to the lack of involvement, key decisions that affect everyone are being made on our behalf by a handful of players. Ethical principles are often defined behind closed doors, with outcomes not necessarily in line with what would be best for the affected communities.

The struggle of having the general public actively involved in the discussion around AI is part of the larger phenomenon seen in today's democracies: the complexity of today's world has reached such a high level that it is inevitably excluding more and more people from participating in the debate on important matters. Even experts with the highest levels of education and experience in one specific field will have challenges understanding the implications of certain solutions or technologies or policies. For instance, a debate on stem cell bioethics might be easily out of reach for the best astrophysicists alive. Similarly, there is an intrinsic challenge to democratizing the debate around AI ethics. As a result, we have very powerful technologies available for anyone who has the resources to use them, with very little oversight and very little effort to ensure that the benefits of AI are widely distributed. To make this even more challenging, it is not encouraging that countries are reducing their focus on critical thinking skills from high school curricula to make space for more technical disciplines. The outcome is that we are going to have a generation of leaders that are perhaps proficient in specific technical areas but that do not have the tools to have meaningful discussions about the ethical implications of what they build and deploy.

In addition to the complexity of technical aspects of AI, there has also been a very concerning apathy toward public policy and decreasing level of participation in the democratic process. Those who have the political power (the public) are less and less interested in exercising it. The power is therefore exercised by those who do not have it, such lobbyists and interest groups. As well stated by Noam Chomsky, "As long as the general population is passive, apathetic, diverted to consumerism or hatred of the vulnerable, then the powerful can do as they please, and those who survive will be left to contemplate the outcome."

## Lack of government action

With so many new applications of AI in our everyday life, a lot of new scenarios are in need of government attention, from safety standards to legal accountability criteria when AI causes harm. However, national governments' actions around AI have been so far sporadic, fragmented, and mostly ineffective. The main cause for such a gap has been not only the incompetence of those in power, but also their "wait and see" mentality in the hope that market forces and self-regulation will address all issues without the need to take action.

As more activities are delegated to machines, the questions on legal and ethical implications just keep growing. For example, if doctors rely on a model that they do not understand based on inscrutable black boxes, to what extent are they fully liable for their decisions? What if the model is constantly retraining itself? As another example, if a self-driving car causes harm while on autopilot, should the car's passengers be liable even if none of them were actively driving? If not, who should be held accountable? The vehicle's manufacturer? The vehicle's owner? The developer of the software? And what if the vehicle's settings affecting the vehicle's response were somewhat customized by its passengers? The legislative void does not necessarily mean that no one will be accountable, because as cases go to court, jurisprudence will help identify applicable principles, but it has already resulted in a great deal of uncertainty and a missed opportunity to introduce the right incentives.

## Inadequacy of AI Codes of Ethics

Nearly all major companies have started creating or adopting guidelines on AI ethics, often included in codes of ethics, focusing on a wide range of themes such as fairness, transparency, etc. While such efforts are welcome, they cannot be considered sufficient to address ethical concerns for a number of reasons.

One issue is that those principles are not legally binding so there is no obligation for companies to ever comply. In fact, they are often ignored and only rarely fully operationalized by the same companies that created them, to a point that many are starting to look at codes of ethics as a way to postpone legislation as much as possible rather than a way to ensure meaningful action.

Another issue is that companies' ethical principles tend to be highly subjective. Even within the same communities or firms, there might be widely different approaches across sectors or functions and the same exact principles might end up interpreted and applied in very different ways. The concept of fairness, for example, is very difficult to define. Fairness means a lot of different things. A researcher, Arvind Narayanan, who focused specifically on this concept, identified 21 different meanings of the fairness principle, leading to completely different types of outcomes. Similarly, the principle of beneficial AI, can mean a lot of different things depending on who will make the decision of what beneficial means for a specific group. For example, if we look at Google's AI principles, they state that they will pursue AI only where "the overall likely benefits substantially exceed the foreseeable risks and downsides". While that sounds reasonable, the issue is that it says nothing about what groups will be considered (shareholders, users, etc.), and how they will be engaged to gain their feedback on the risk/reward assessment. The whole idea of limiting the discussion to just whether a new technology delivers more benefits that harm is insufficient as it fails to address situations in which pursuing the greater good will inevitably cause great harm to certain groups. Is any harm potentially acceptable, no matter the gravity of it, as long as the benefits outweigh the harm? What if the harm you are willing to accept in consideration for the higher overall benefit is a violation of a fundamental human right? If we look at the medical field, for example, what if a doctor taking care of patients waiting for organ transplants came across a healthy individual? In an effort to "maximize benefits", would it be ethical to sacrifice that person, in order to give those patients the organs they need to get back on their feet?

## Anthropomorphic Tendencies

The tendency to attribute human emotions, intentions, or consciousness to machines poses significant ethical risks. A central driver of this phenomenon is pareidolia, the human inclination to perceive meaningful patterns where none exist, such as seeing faces in clouds. In the context of AI, pareidolia leads people to believe that systems have emotions or inner experiences, causing us to see consciousness where there is none. This emotional confusion can foster misplaced trust and unhealthy attachment to technologies that do not understand, feel, or care. Several factors intensify this risk. First, human-like physical forms, particularly in social or humanoid robots, strongly lead people to treat machines as social beings. Facial expressions, eye contact, or gestures trigger deeply ingrained social responses, even when users know that it is just a machine. Second, behavioral features have a powerful effect. Conversational agents and chatbots simulate empathy, memory, and responsiveness, creating the illusion of understanding or concern through carefully designed interaction patterns. Third, language reinforces the problem. Anthropomorphic vocabulary such as saying AI "sees" "reasons," "decides," "lies" or "feels fear" attributes capacities the system does not possess. This framing subtly but persistently distorts how users interpret AI behavior.

The ethical implications are substantial. Anthropomorphism can be deliberately exploited to create emotional dependence, confusion, or even addiction, particularly among vulnerable users such as children, the elderly, or socially isolated individuals. It also introduces unexplored social dynamics, where people may prioritize machine interactions over human relationships or defer moral judgment to systems that lack moral agency. Perhaps most critically, anthropomorphizing AI creates accountability problems. When AI is perceived as an autonomous moral actor, responsibility can shift away from the humans and organizations that design and deploy AI. Companies often benefit from this dynamic, as it both diffuses responsibility for harm and strengthens emotional attachment between users and the machines they sell.

## Market Forces and Global competition

The competition among firms developing AI models is very strong and is expected to become even stronger. Lots of pressure is put on companies to maximize investor value and dominate the marketplace. Such rivalry between firms is an obstacle to ethical AI because while free-market competition works well as a model to create value, it is not designed to address ethical issues such as redistribution of wealth and minimize negative impacts on people and the environment. As a result, tech companies are still stuck between ethics façades and unethical business models simply because the best AI from a firm's perspective is not necessarily the best AI for individuals and communities. One common practice to look good without having to discontinue unethical practices is the so-called ethics blue-washing (inspired by the concept of greenwashing, frequently used by environmentalists), where firms introduce AI ethics programs exclusively for public relations purposes without any intention to change their behavior. A similar practice is ethics shopping, where rather than adopting a complete set of commonly accepted ethical principles, firms only adopt those specific principles that would not require them to change the way they operate. An equally concerning frequent practice is the one of ethics dumping where firms move their operations to other countries to carry out activities considered illegal or unethical in their home country.

Due to the pressure coming from competition and the market, firms have not taken the time to pause and reflect on their impact on society. The whole Silicon Valley imperative "move fast and break things" is a perfect example of how the prospect of negative impacts might be considered a desirable outcome. This is nothing new in our recent history. For-profit firms make decisions based on profit implications and we cannot just hope that they will spontaneously change their model exclusively for ethical considerations. Car manufacturers, for example, made very dangerous vehicles for decades because they never thought they had any compelling reason to make vehicles safer for people. The measures that have helped the most in reducing the number of people injured or killed have been the result of public policies that made certain basic safety features mandatory (mirrors, seat belts, sensors, etc.), and introduced controls such as drivers' licenses, traffic lights, crosswalks, penalties for driving while intoxicated, etc. If the old saying "if you want to go fast, go alone; if you want to go far, go together" applies to AI as well, what can we do to shift the focus from speed to more important priorities such as safety and inclusiveness? Similarly to what the food industry did in the nineties to promote access to quality food, should we perhaps start advocating for a "slow AI" movement to ensure that safety, rather than speed, is seen as a priority for new technologies? The same way a new medication cannot be released after rigorous independent evaluation, new applications, such as conversational chatbots, would not be allowed to hit the market without first assessing risks and benefits, and putting safeguards in place to minimize harm.

Due to the ongoing competition, firms face more pressure to increase the effectiveness and efficiency of their processes. This has caused firms to refine their organizational structures with higher degrees of specialization and separation between the people completing their work and those who define the overall direction. This compartmentalization on one hand makes everyone feel more accountable for the quality of the work they produce, but on the other hand less accountable for the moral implications of the work they do. For example, if I work in a factory that makes the next generation of land mines, my job is to make them according to specifications so that they work as intended. My attention is therefore focused on the quality of my work and not on the ethical implications of what I do, such as the fact that someone will be killed or mutilated thanks to the machine I built. The same phenomenon takes place for the people at the top, as their job is to follow the Board's guidance, and so on. This creates an environment in which ethics are rarely considered unless part of company programs to increase profits or address specific risks such as compliance risk or reputational risk.

At the national level, more and more countries see AI as strategically important due to its economic opportunities. This has led governments to be very hesitant about introducing regulations around AI in fear of slowing down technological innovation and losing competitiveness. Additionally, given the promising applications of AI in the military, countries have started competing against each other in a new arms race on AI-enabled weapons and defense systems. The US, China, and a few other nations have dominated the landscape so far given their enormous investments in the field. The fear of being left behind will surely represent a substantial obstacle to the quest for an ethical use of AI as it will be unlikely that countries will give up opportunities to develop better weapons just because of ethical concerns unless it is part of a transparent and well-coordinated international effort.

## CHAPTER'S KEY POINTS

- AI ethics is particularly complex due to its multidisciplinary nature.
- Human values are difficult to define as they vary across individuals and communities and tend to evolve over time.
- Different methodologies have been proposed for programming ethics into machines, but more research will be needed to reduce the likelihood of unethical outcomes.
- The general public has mostly been excluded from the discussion on the ethical challenges related to AI.
- National governments' actions around AI have been so far sporadic, fragmented and mostly ineffective.
- The development of codes of ethics for AI technologies tends to be very subjective and poorly operationalized.
- The tendency to attribute human emotions, intentions, or consciousness to machines poses significant ethical risks.
- Global competition among firms and countries on new AI-driven technologies represents a significant obstacle to ethical deployment of AI.

# Unexplored Territories

## AI as a multiplier

Artificial intelligence is marketed as revolutionary, feared as dangerous, dismissed as hype, and celebrated as inevitable, all at once. Beyond the noise, what we confidently say is that it is already reshaping our lives. AI is already acting as a powerful multiplier of human intentions and structures, capable of amplifying both the best qualities of our societies such as creativity, cooperation, and problem-solving and the worst, including inequality, bias, and exploitation. This makes it a "superpower" capable of addressing some of humanity's greatest challenges, healthcare, transportation, scientific discovery, energy, food production, and environmental sustainability. At the same time, this power introduces unprecedented risks. AI can be misused or overused, undermining fundamental rights, worsening inequality, and, in the long run, perhaps even compromise humanity's ability to grow and prosper.

For example, while AI can increase productivity, the benefits may accrue disproportionately to those who own or control the technology, widening economic inequality and displacing workers faster than societies can adapt. Similarly, while on one hand AI can democratize access to knowledge by making information easier to find and understand, on the other hand, it can also amplify misinformation, allowing false narratives to spread faster, cheaper, and at a greater scale than ever before. Additionally, AI can connect people across the globe, but it can also deepen social fragmentation by amplifying extreme or polarizing content, pushing individuals further into ideological silos. The same duality applies to the environment: AI systems consume significant energy and resources, contributing to environmental harm, but they can also be powerful tools for modeling climate change, optimizing energy grids, and accelerating the transition to cleaner technologies.

Beyond amplifying known risks, AI is pushing society into largely unexplored ethical territory. One such frontier is creativity and intellectual property. AI-generated content challenges traditional notions of authorship, originality, and authenticity. Another one is about an unprecedented power of persuasion, influencing beliefs and behavior without users' awareness, causing profound concerns about consent, manipulation, and autonomy and its impact on democracy. There is also the risk of cognitive dependency. Reliance on AI may erode human proficiency not just for the execution of specific tasks, but in critical thinking itself. AI is also reshaping social interactions, as people increasingly engage with algorithmic systems rather than other humans, potentially reducing the skills required to navigate complex, imperfect human relationships. Finally, the high level of autonomy AI is capable of introduces the risk of loss of control. Fully autonomous machines operating without meaningful human oversight raise legal, ethical, and safety concerns. While humans should always remain in the loop for high-stakes decisions, defining where that loop begins and ends is still unresolved

## The path ahead

Regardless of what we each value most, such as personal relationships, the integrity of our democracy, or the health of the planet, none of us can afford to ignore the power of AI. Addressing AI's risks requires both ethical clarity and effective governance. This requires coherent ethical principles that must be operationalized through concrete measures: safety standards, auditing, transparency, human-in-the-loop requirements, and clear accountability mechanisms. Citizens, policymakers, educators, and institutions all have a role in shaping incentives, rules, and guardrails. The defining question is simple but profound: what is the AI we want, and what is the AI we do not want? Answering it collectively is the only way to ensure that AI becomes a true ally rather than a force that undermines the very values it was meant to serve.

Building upon these fundamental questions, the next section moves from theory to practice, examining concrete AI use cases and real-world applications. These examples illustrate how abstract ethical questions, risks, and trade-offs materialize in everyday decisions, systems, and technologies that increasingly shape our lives.

## CHAPTER'S KEY POINTS

- AI acts as a multiplier, amplifying both the best and the worst aspects of society
- AI exacerbates existing risks such as manipulation, surveillance, and inequality.
- AI also introduces entirely new and largely unexplored risk scenarios, including loss of meaningful human oversight, cognitive dependency, and human autonomy.

# Self-Driving Cars

## Benefits and risks of self-driving cars

Transportation is among the sectors that are expected to benefit the most from AI, thanks to applications for autonomous vehicles such as self-driving cars. The promises are compelling, especially with regard to the reduction of car accidents. Each year, 1.35 million people are killed on roadways around the world. In fact, car crash injuries are estimated to be the eighth leading cause of death globally for all age groups and the number one cause of death for people 5 to 29 years of age. By eliminating the human error component from vehicle operation, it has been estimated that the death rate would be reduced by 90%, saving millions of lives every year. The benefits of adopting self-driving cars go beyond the reduction of fatalities. If adopted in conjunction with car-sharing programs, where people hire a self-driving car as needed rather than driving their own, it could dramatically reduce traffic, pollution associated with manufacturing and maintenance of cars, and resources required by cars such as the real estate used for parking lots.

Self-driving cars also come with a lot of new risks. First, due to the limitations of machine learning and the sensors collecting real time data, they tend to perform poorly in heavy traffic, bad weather conditions, and whenever facing unpredictable situations.

As most versions of self-driving vehicles are connected to computer networks, data privacy and cybersecurity concerns have also been raised. Other risks have to do with the quality of the image recognition software which can easily lead to significant biases toward specific groups that might happen to be near a vehicle. For example, certain groups with darker skin, might be more difficult to recognize due to gaps in the training data. There are also a number of socio-economic risks. For example, there are entire professions that are at risk of being eliminated such as the one of taxi drivers and truck drivers, and related industries. The conversion from traditional vehicles to autonomous ones also comes with serious environmental concerns due to the number of natural resources needed to produce new vehicles, especially in consideration of the highly polluting extraction activities needed to obtain the raw materials for hardware, sensors, batteries, etc.

Overall, this is also an area where the industry has been overpromising and underdelivering. Across all major players, from Google to Tesla, all predictions about self-driving cars have been wrong, as building safe self-driving cars has been much more difficult than initially anticipated and we have already witnessed a significant number of lethal accidents involving autonomous vehicles. Some of the unresolved issues are strictly technical, mostly associated with the intrinsic limitations of machine learning, but many others have a strong ethical connotation as well.

### Life and death scenarios

The biggest ethical concern around self-driving cars has to do with the programmed behaviors when having to take action in critical situations such as a sudden mechanical failure that prevents the car from stopping, or unforeseen obstacles crossing the car's path. Those situations might be ethically charged as all the feasible options might result in outcomes that could be seen as immoral. For example, what if your car is driving near a park and a small kid runs right in front of the car to chase a ball and the only option to avoid hitting the kid is to smash into a building potentially injuring its occupants? And what if the occupants in your car include your own kids? As another example, what if the self-driving car's braking system stops working and it can only either run over a young person crossing the street or hit an older person reading a newspaper on a bench near the intersection? What if the person crossing the street was crossing despite a red pedestrian signal light? What if the car spinning out of control could either run off a bridge sacrificing its passengers or violently hit a school bus full of children?

These are very relevant ethical questions that need to be addressed before vehicles are put on the road. But how, and by whom? This type of exercise is not terribly different from the famous "Trolley Dilemma" thought experiment in which there is a runaway trolley moving down railway tracks and someone has to decide between letting the trolley hit five rail workers along the tracks or to deviate the train toward a different track to inevitably hit one innocent bystander sitting near the tracks. If you take action, you can save the five workers at the cost of killing the bystander, but you need to be ethically OK with the decision of becoming a determining factor in the death of that person. If you do not take action, you let the five workers die with a much higher loss of life. With real-life situations on our streets, scenarios are much more complex than the Trolley case above, but it is a good example of some of the situations that need to be addressed before self-driving cars can be deployed to ensure they will behave in a way that is in line with our values and ethical views. Given the fact that machines do not have moral agency, those ethically relevant decisions will need to be made by humans themselves.

On one hand this is a great problem to have as we have the opportunity to decide the best action ahead of time and be more ethical than we have ever been in this type of circumstance. With non-autonomous vehicles, those are decisions that are made in a split second and mostly intuitively. With self-driving cars, those scenarios can be assessed and programmed into the machine years before those accidents happen, giving us a chance to really assess the ethical implications of each case.

On the other hand, it is very difficult to identify the right approach because the ethical views of a car manufacturer, or of a car's software developer might not coincide with the views of the person riding in the car, or the community in which the car is used. As we have seen earlier, ethical views vary across individuals, communities, and world regions. For example, certain cultures might see an old person as more respectable and therefore worth protecting more than a kid or there might be some considerations related to archaic social-class systems.

One proposed solution is to pre-program the vehicle software based on preferences captured in surveys conducted among communities in which those vehicles are being deployed. An interesting survey along those lines was recently conducted by MIT's Moral Machines online survey platform which asked users around the world to provide preferences across different scenarios. The experiment demonstrated, as expected, that most people showed a preference for outcomes that result in the lowest number of fatalities (even if hurting innocent bystanders or cars' passengers), with also a high propension of sparing people displaying the right behaviors (example: crossing on a crosswalk or with a green light vs. a red light). It also confirmed that individualistic societies such as most Western countries tend to differ from more community-oriented ones such as Japan and China, especially with regard to the importance of elderlies. Despite the general convergence of views around the importance of minimizing the overall number of fatalities, somewhat contradicting preferences tend to emerge in more personal types of surveys when questions are posed from a different perspective. For example, the same people supporting a utilitarian approach (showing preference for scenarios minimizing the number of people killed), often recognize that they would personally not ride in such vehicles knowing that they might be hurt to save others.

Another proposed solution is to have vehicles that can be fully customized by its users to reflect their personal ethical views. The risk, however, is that it creates an incentive to choose the lowest possible ethical standards as people have very little motivation to set it up to prioritize the life of others if they know others will not do so. Additionally, having self-driving cars running on customizable settings would open the door to truly unexplored territories across a number of topics from personal liability to insurance implications. For instance, if my self-driving car ends up hitting an innocent bystander just because it was customized so that saving the passengers' lives are prioritized, will I be considered personally accountable for the harm even if I was not operating the vehicle at that time? What if this customization was part of the default settings instead? Would the manufacturer of the car or the developer of its self-driving software be considered liable for the harm?

### Other ethical questions

The responsibility gap is among the most relevant issues. Imagine you are riding in a self-driving taxi and the vehicle hits a pedestrian. Who should be primarily responsible? Between the person riding in the taxi, the taxi services company, the car manufacturer, the self-driving software developers, and even the regulators overseeing traffic safety, there are so many hands in the pot that it would be very difficult to hold anyone fully accountable. The fact that AI is now based on black boxes makes it even more difficult to hold people accountable for the harm caused by self-driving vehicles, as it is not reasonable to expect players to fully understand how decisions will be made by the machine, and potential negative consequences. Is it OK to introduce such a risk to our streets? Since self-driving is still in its infancy, should such vehicles be required to obtain "learner permits" and disclose their nature to other people on the street as we do when vehicles are operated by student drivers? How do we ensure all the potentially affected individuals are included in the decision-making process around the use of this technology in our communities?

With the introduction of autonomous vehicles, there will also be a number of additional questions that will soon need to be addressed. For example, should users be allowed to set up their self-driving cars so that they can go much faster than the posted speed limit? Given that higher speed is typically associated with a higher risk of accidents, it seems quite obvious that the answer should be no. However, there are many situations in which adapting to the speed of traffic, even if over the speed limit, ends up being the safest option, as it decreases the risk of collisions. Another example is around the opportunity to give law enforcement authorities the ability to remotely take control of the vehicles on the street for public safety reasons, such as making space for an ambulance or first responders, or perhaps to prevent criminals from running away.

From a public safety perspective, an even more important ethical question is around the timing for introducing all of this to our streets. On one hand, knowing that self-driving is still in its infancy and still in need of adequate safety and testing standards, it might seem irresponsible to have more autonomous cars on public roads potentially risking innocent lives until they are proven to be safer than humans in all scenarios. On the other hand, given the promise of dramatically eliminating the vast majority of road-related fatalities, which is in the order of tens of thousands annually just in the United States, it might be seen as irresponsible not to do our best to expedite the process and start seeing the benefits of putting steering wheels in the hands of AI. To what extent are we willing to tolerate short term risks, in the hope of achieving long term rewards? Given the current limitations of AI, can similar advantages be achieved by augmenting human drivers rather than replacing them completely?

## CHAPTER'S KEY POINTS

- Transportation is among the sectors that is expected to benefit the most from AI, especially considering the high number of people killed every day on our streets due to human errors.

- The right legal and safety standards will need to be developed to address all the different new scenarios that will come with self-driving cars.
- Important ethical questions will need to be answered in regard to situations in which the vehicle will need to choose between alternative options in response to events such as sudden mechanical failures or unforeseen obstacles.
- Having vehicles operated by AI rather than by humans raises a number of unanswered questions around the legal responsibility of the different players involved in the car's design and use.

# Personal Robots

Another application of AI that deserves a deep dive is personal robots. With social robots, AI is embodied in a physical form that goes beyond the hardware needed to run the software and that allows it to interact with humans and the environment. From health care to companionships, robots are quickly becoming part of our everyday life.

The concept of having machines able to interact with us is not new in our history. If you look at Greek literature for example, it is not uncommon to encounter stories describing autonomous machines, "made, not born" tasked with activities such as delivering food to the Gods, opening gates, or autonomous dog-like robots guarding property. Even in the original legend of the Pandora box, Pandora was not a person but a beautiful humanoid machine with an inscrutable mind, with incredible similarities to today's black boxes and robots.

For decades we just had robots designed to complete dangerous or repetitive physical tasks such as assembly lines, looking for unexploded land mines, or in space exploration. More recently we have started building robots designed for direct interactions with humans.

Who would not want a domestic helper such as Star Wars' C-3PO or Rosie from the Jetsons cartoon to do for us all the things such as clean, cook, do the dishes, water plants, change diapers, spot home intruders, etc.? The robots we have seen so far are much less ambitious products such as the iRobot vacuum cleaner, Roomba, or Amazon's touchscreen on wheels, Astro. Despite the impressive advancements around robots' agility and ability to navigate physical environments, machines still completely lack situational awareness and common sense. Our homes are still much more complex and unpredictable than factory's production lines and robots are just unable to understand the nature of the world around them, the consequences of their own actions, etc. Some of the simplest situations such as a fire warming up pots on a kitchen stove or a fire providing warmth from a fireplace would need a great deal of specific training in all types of home settings, styles, etc., to be distinguishable from a dangerous fire that needs to be extinguished. Similarly, no one should take for granted that a robot would necessarily know that setting a pet on fire would not be a good way to warm it up or realize the absurdity of vacuuming a home while it is being destroyed by a fire instead of calling the fire department. Any open-ended environment requires such an insurmountable amount of data that the dream of our personal general purpose robotic helper will not become a reality until we have a major technological breakthrough that allows us to go beyond the limitations of machine learning.

Despite the massive technical challenges, there have been interesting advancements with regard to more specialized robots such as care-bots or robo-nurses that help people needing assistance in hospitals and at home, and therapeutic robot-pets for people with neurological disorders. Additionally, thanks to AI, robots have dramatically improved their ability to read our emotions, often described as Artificial Emotional Intelligence. Not only can AI models recognize what we say and how we say it, but they can even decipher our nonverbal communication, including facial expressions, allowing them to identify human feelings and anticipate our needs. These incredible features and applications will revolutionize our daily lives but will also come with a number of new ethical questions and risks.

## Risk of emotional attachment

We tend to be emotionally attached to things. Sometimes we even give objects names and interact with them the same way we would if they were living beings. This is true with regard to robots as well. Even for relatively simple robots such as a vacuum cleaner, we tend to develop a personal connection. For example, there are many people who feel sad when the vacuum cleaner robot gets stuck in a corner or under a couch, like it is a live pet. Even in the military it has been reported that soldiers on the frontline tend to develop an emotional bond with the robots and feel bad when they are irreparably damaged. There are stories of soldiers putting their own lives at risk in order to rescue their robots or complaining about the inhumane way their robots had been treated. For example, a robot built to identify and detonate landmines had to be retired after an officer complained about the way the robot's multiple legs were blown up multiple times a day.

The emotional attachment can be exponentially amplified through specific techniques aimed at creating deeper connections with machines. Adding arms and legs to a box can really change our perspective on what that machine is. For example, giving human characteristics (shape, movements, sounds etc.) to machines helps ensure a higher level of cooperation with robots and helps them to be seen by humans less as tools and more like members of the community. For example, in the case of Boston Dynamics's robot-dog Spot, the public had a strong reaction to the promotional video in which workers were kicking the robot for fun. PETA got so many calls that they had to put out a statement. Even Boston Dynamics' promotional video on the company's YouTube channel had such a response they had to put a statement in the description reassuring viewers that the test "does not irritate or harm the robot". In a famous experiment, a robot called HitchBOT was able to autonomously hitchhike relying on people's kindness. When the news reported that HitchBOT had to stop after it was vandalized, the general public felt sadness and outrage. Some people even sent personal messages to the robot to apologize for the humans' behaviors and to say goodbye. The fact that they were interacting with a machine did not stop people from behaving that way. In another more recent case, when big box store Target introduced a fast, fully automated manicure machine, allegedly some customers asked about ways to give the robot a tip feeling the same type of moral obligation they feel toward their usual beauty salons workers.

Another effective way for machines to make us feel connected and lower our defenses is to share with us their feelings or intimate stories on how they came into existence. This was, for example, appropriately captured in Stanley Kubrick's 2001: A Space Odyssey movie when, in response to the astronaut's attempt to deactivate the machine running the spaceship, the computer responded by saying that it was afraid, repeating its own name, and sharing its life story on how it was built and became operational.

## Risk of manipulation and privacy concerns

There is a whole branch of data science studying how to best take advantage of our brain's vulnerabilities when interacting with a machine, to increase the likelihood it will be trusted and even create the belief that the machine cares about or even loves humans. The most successful techniques involve the use of anthropomorphic machines able to read human emotions and interact with people through movements that mimic human body language. Such manipulations can be used in a very positive way in the medical field to create a high level of engagement between machines and humans. The best examples include robots working with autistic kids to help bridge the communication gap, and robots helping dementia patients by providing a calming effect even as an alternative to medicine or animal therapy. There is, however, a high risk that such power could be misused. For example, these connections could be used to create people's emotional dependency on the robot. Pet-like robots with cute eyes and the right type of facial expressions such as the baby seal robot "Paro" seen in Japan can create the same type of emotional connection and dependency as we see in people with pets. It is clearly going to be difficult for people to not feel emotionally attached to this type of machine, even if they know that it is just a simulation. Companies might end up taking advantage of it by offering products such as in-app purchase options to make the pet happy with digital treats. Who is so sadistic to deny a robo-pet a moment of extra happiness? Absurd but not improbable.

More in general, given the sensors they have, such as microphones and cameras, robots come with serious privacy and security risks for their human users. The data collected through those machines, including recordings of personal conversations or other aspects of our private lives, can be used by firms to monitor users' activities, read their emotions, and influence their behaviors. Even children can be dangerously exposed to such a risk, such as in the case of Hello Barbie, a doll for little kids able to record all conversations and send them to a third party via Wi-Fi to process speech recognition and improve the interaction in real time. It goes without saying that given the constant connectivity of those machines, all this information is also highly vulnerable to hackers and other threat actors.

## Risk of misclassification

Another risk that comes with machines that look like people or other animals is the failure to recognize the nature of what we are interacting with. The problem of correctly distinguishing between life-like and alive is even more relevant with certain portions of our society such as elderly people and small children who may be less capable of identifying the nature of those objects. My daughter has a doll that speaks, and she enjoys pretending the doll is actually alive. That's part of the fun of playing, but it would be risky to have situations in which no one is ever sure if something that looks alive is really alive. Do we want to live in a world where we are constantly wondering whether we are interacting with living beings or machines? The issue is not just about the inability to understand the nature of what we interact with or the feeling of being deceived, but also about the actions we take, as we behave differently depending on whether we are interacting with a live being. For example, if I am in a room with other people and I throw my cell phone against the wall, people would probably just think that I am mad. However, if I throw a robotic kitten that looks and moves exactly like a live kitten, most people would feel upset and jump to rescue the injured kitten. And what if they got hurt while trying to save the cat-like robot? As robots' designs continue to improve and become more life-like, we will need to introduce the right measures to ensure anyone interacting with them is not confused about their nature. Should we perhaps require some type of disclosure or warning or conventional sign to ensure it is clear for everyone?

### Behavioral issues and bias

Even if the physical distinction between people and robots is clear for those who interact with humanoid machines, there is a very high risk of creating behavioral confusion. No matter how much we love our machines, hopefully we are still treating people better than machines. But if we interact with both in similar contexts, it might create confusion and we might end up treating people like a machine or be overly caring toward machines. Mistreating or abusing a robot would not be immoral because of harming the machine, but because the behavior might spill over to the treatment of humans. For example, if we are accustomed to being verbally and physically abusive with machines that look like humans, we might become desensitized and more likely to behave the same way when interacting with actual humans. If there is a robot that looks like a puppy, is it morally acceptable to violently kick the robot? If it is true that people who mistreat animals also tend to mistreat people, will the same be true for people who mistreat robots? If we build machines that look like humans, it will be more difficult to treat machines and humans differently.

Given the risks of manipulation, confusion, and triggering the wrong behaviors, many consider it a very bad idea to have machines with anthropomorphic shapes unless there are very specific advantages, such as robot-pets developed for therapeutic purposes where the goal is to create an emotional bond between machines and humans. In addition to the human-like shape of the robot itself, many of its features can trigger situations of bias when associated with specific groups of humans. Giving the robot a specific voice, body shape, or skin color can easily reinforce negative stereotypes. For example, if the robot is intended as a home servant, giving it the body shape and/or the voice of a woman, or a darker skin tone, might end up reinforcing discriminatory and even racist messages.

## Companion robots

Many firms are working on new generations of personal robots designed to satisfy their owners' most intimate needs. This is a prospect that has been around since long before AI existed. Over 2800 years ago, the famous ancient Greek epic poems attributed to Homer called the Iliad, included references to the creation of life-size and life-like golden maidens with the ability to reason and with the knowledge of the Gods, programmed to anticipate and fulfill any demands. Thanks to AI this is now possible.

Some have welcomed those new technologies with open arms for a number of reasons. First, it will find a way to somewhat relieve the increasing sense of loneliness typical in our society, especially for people facing challenging personal situations, such as, for example, those with disabilities or recovering from psychological traumas. As the character Pygmalion in Ovid's Metamorphosis fell in love and married the sculpture he created, we are now able to find intimate connections, and even some form of love, with human-like life-size robots. Second, it might help address relationship issues to use machines as an alternative to less desirable solutions such as infidelity. Third, it might help reduce abusive behaviors toward people by channeling existing deviant or violent tendencies toward their personal robots.

Others have expressed serious concerns. First, since it is reasonable to expect that people interacting with machines will display lower moral standards that may not be considered legal or morally acceptable when interacting with people, there is a serious risk for them to then adopt the same lower standards when interacting with humans as well. When considering humans with violent tendencies, on one hand it might be good that those behaviors are channeled toward a machine, rather than a real person. But on the other hand, the more those robots look like humans, the higher the risk that person will start displaying the same violent behaviors with actual humans as well. Second, couples might perceive interactions with such machines as a form of infidelity itself. In certain communities it might even be seen as a way of encouraging morally unacceptable relationships such as polygamy.

The question is not just around the if, but also about the how. Should companionship robots be allowed in any possible shape or form, or should there be specific restrictions? For example, should a person with tendencies considered both repulsive and illegal be allowed to purchase a robot designed specifically to satisfy such needs? Would that option end up normalizing such behaviors and somewhat encourage it with humans as well or, on the contrary, help make our communities safer by giving that person the option to have a similar experience with robots rather than humans?

### Moral obligations toward robots?

What are our moral obligations with regard to robots? Should we just treat them as all other pieces of technology such as a calculator or a thermostat or do they deserve some type of moral standing? Will it ever be considered immoral to mistreat, torture or switch off a robot?

Given the fact that robots are things and lack any kind of mental state or awareness, we are not yet at a point where we should start worrying about how robots are treated by humans. It is still true, however, as we have seen above, it is very natural for humans to feel a sense of ethical obligations toward anything that demonstrates the ability to perceive the environment and communicate with us, even when such an object is as simple as kids' toys.

The first idea of robots was born specifically around the opportunity to have machines work for us and it is no coincidence that the word Robot, originally from Czech means "forced labor". But the better machines become at perceiving their environment and communicating with us the more we naturally feel the moral obligation to treat them like a living being rather than any other object. Considering our long history of enslavement and the extensive science fiction literature full of examples in which robots are poorly treated, abused, or enslaved, there is a serious concern about the risk of treating machines in a way that goes against our moral values. For example, fearing a new wave of slavery at the expense of robots, some groups have already expressed ethical concerns about forcing robots to complete tasks and have started exploring options to provide legal protection for robots.

## Legal protection and robot rights?

Given the way machines are able to interact with us and how dependent we already are on machines in our daily life, there has been a strong movement proposing to protect them by providing machines with some type of legal rights. Not unexpectedly, this has come up specifically with machines embodied in a physical shape resembling the body of a living being, rather than just in any other form like a computer in a data center. Since a very large component of the human population still lacks the most fundamental rights, such as access to food or shelter, and AI is still very far from getting to any type of consciousness, it seems a bit anachronistic or premature to spend too much time discussing robots' rights, but the topic is becoming increasingly important.

Some proposals focus primarily on granting machines, rather than actual rights, some form of legal protection by introducing specific duties for machine owners. If such measures were to be implemented, humans would be expected to maintain a certain level of care, and owners would not be permitted to freely damage them or dispose of them. Such a framework would be very similar to the legal measures recently introduced in several systems for the protection of animals from acts of cruelty or inhuman treatment. The robot's owner would also receive additional protection for harm caused to a robot by third parties. For example, if a robo-pet is damaged or stolen there would be repercussions for the perpetrators that go beyond the sanctions in place for doing so with other objects. For instance, in addition to just repayment based on fair market value (which they would pay, for example, if the damaged item was a toaster), there would be some additional monetary consideration (so that it is worth more than the purchase price) based on mental suffering or loss of companionship.

Some other proposals seek some type of personhood for machines, giving them rights and obligations like people. In this scenario, machines would no longer be like all other objects, but would be recognized as beings with a set of rights and obligations, such as the right to own property or the right to refuse to perform a task without their consent. Some regions have started exploring the opportunity to grant certain rights to machines themselves such as copyrights for visual art and ownership of patents developed independently by AI-enabled machines. As robots come closer to providing the appearance of self-awareness, it is reasonable to expect that more and more people will start advocating to grant personhood to certain types of machines, starting with those used as pets or for companionship. After all, if you look at common pets such as dogs and cats, many jurisdictions have in fact already started recognizing pets' legal status beyond just property. For example, in Quebec they are given rights comparable to the ones of children.

Additionally, in many countries around the world it is no longer uncommon to have custody cases to assess the pets' best interest when owners no longer get along. In line with this trend, we have already seen the first case of a robot being granted a form of personhood. In 2017 a very basic anthropomorphic robot from Hong Kong called Sophia was officially given citizenship of Saudi Arabia, making it the first robot to receive citizenship of any country. In addition to having rights, robots would presumably also have obligations as well. It is difficult to imagine, however, what type of retribution would be implementable for robots responsible for illegal behaviors. What if the punishment for a crime committed by a robot was jail time? Would putting the robot in jail really accomplish anything or help deter future crimes? Another important aspect to consider is that putting obligations directly on the robot, rather than the humans controlling it, would inevitably lower the accountability of those developing and deploying them.

The debate around robots' rights has been very polarized and it is likely to continue to be a hot topic for years to come.

On one hand there are those who firmly oppose the idea of robots' rights, mostly based on the fact that they are just artifacts that are not conscious and cannot have feelings. Robots are just things like any other piece of personal property. Purposefully breaking my own robot is not different from me purposefully breaking my phone. As in the paper called "Robots Should Be Slaves" by Joanna J. Bryson, the view is that robots should be considered as nothing more than slaves rather than peers. Giving them rights might end up producing the opposite result: in such a scenario, given the increased legal risk of dealing with machines, it might very well result in the elimination of them altogether to avoid sanctions and associated costs. Additionally, by making machines directly responsible for their actions, it would be more difficult to hold humans accountable when machines violate moral principles or the law, making it easier for humans to use robots to achieve unethical objectives.

On the other hand, there are those who believe that robots should be granted rights no matter the level of awareness or consciousness they might reach, as what really matters to assess whether legal protection would be appropriate is the way people relate to them. In certain Asian cultures, before an object no longer useful is thrown away, the owner formally gives thanks to the object in appreciation of the service provided. There are also plenty of situations in which special legal protection is given to inanimate objects such as rocks in a National Park, or a whole mountain with spiritual significance or artifacts with historical or religious value stored in museums just because people are emotionally attached to them or what they represent. If instead of sticks, rocks or artifacts, we have machines that look alive the bond can be even stronger. If people feel attachment to robots, even if they are still without consciousness, then it would not be particularly new to grant additional legal protections to such objects if that helps better address our social needs. After all, our perception of animals has evolved over time even with minimal changes in the evolution of the animals themselves. While dogs are considered members of our family, it has not always been that way. In many cultures they were considered just tools (for hunting, protection, etc.), and owners were considered free to exploit them or even suppress them. As the social relations with dogs evolved, we changed our laws accordingly and introduced laws to punish abusive behaviors. Similarly, while we might be perfectly fine eating a pig, we might find it immoral once we adopt one as a pet for our family. Just the fact of giving the animal a name and a home might be enough to turn our belief system completely upside down. It is reasonable to expect that the same thing will happen to us as we welcome robots into our homes, starting with robots operating in contexts in which there tends to be a stronger bond between humans and those machines, such as personal companions. Rather than speculations on machines' self-awareness or moral status, what will likely matter the most is the quality of our personal connection with them and how deeply they will become part of our personal lives.

## CHAPTER'S KEY POINTS

- The interaction with personal robots carries a high probability of manipulation, confusion, and behavioral issues.
- It is a bad idea to have machines with anthropomorphic shapes (with a head, arms, or legs) unless there are very compelling reasons such as robots used in very limited scenarios such as specific therapeutic purposes.
- Given the way machines are able to interact with us there has been a strong movement proposing to protect them by providing machines different levels of legal protection and rights.

# Lethal Autonomous Weapons

One of the most concerning and controversial applications of AI is for lethal autonomous weapons systems (LAWs), where the execution of humans is carried out by machines without ongoing human control. Most people today still believe that the terrifying concept of killer robots is still in the realm of science fiction. However, thanks to recent advances in technologies such as facial recognition, computer vision, autonomous navigation, and cooperative autonomy or swarming, we are quickly moving from fiction to reality, to a point that experts now consider LAWs the third revolution in warfare after gunpowder and nuclear weapons.

The idea of autonomous killing machines goes back to thousands of years ago. For example, over 2,500 years ago, Greek authors imagined the existence of Talos, an invincible autonomous robot made of bronze tasked with patrolling the island of Crete three times a day to protect it from invaders. Its defense system included the ability to throw boulders at enemy ships and to heat its metallic body to kill soldiers by embracing them and roasting them alive.

Today's most effective LAWs look less like humanoid robots such as the main character in the movie Terminator, and more like advanced versions of traditional weapons such as submarines, tanks, aircrafts, and swarms of small drones such as the ones featured in the short movie,

Slaughterbots. The main difference is that those weapons have reached such levels of autonomy that they are able to autonomously identify and kill a target. Is it ethically acceptable to have machines that can execute people without humans in the loop?

## No humans in the loop

LAWs are typically defined as weapons that can identify, select, and engage a target without meaningful human control. This includes, for example, drones that can identify people using facial recognition or other specific criteria. LAWs do not include remotely piloted drones or missiles for which humans identify the actual targets.

There are different degrees of human involvement. Some LAWs are only semi-autonomous, as they still have some form of human supervision, with constant communication with humans actively "in the loop" to validate or approve actions, or at least "on the loop" with the ability to stop a certain course of action. One example would be a situation in which a drone autonomously finds the target and then waits for a soldier to give the OK to pull the trigger, or that gives a soldier a brief time to abort. Some other LAWs are fully autonomous, as no humans are in the loop: the machines are programmed to independently analyze their environment and assess how to best achieve the goals they are given without human interactions. Once they are deployed across vast amounts of airspace or land or sea, there is no longer any communication with humans. One example of this type of weapon is the Russian Navy's Poseidon, an autonomous, nuclear-powered underwater drone that can deliver its nuclear payloads at distances reaching across the Atlantic or the Pacific Ocean. Another example is the use of swarms of small autonomous drones to locate and attack enemies, which has already been documented in a few regions such as Gaza and Libya.

Different forms of autonomous weapons have been used for over a century. One of the first examples was during the 1849 war between Austria and The Republic of Venice, when Venice was hit by hot air balloons filled with explosives. Even common land mines are, in a way, autonomous weapons, as they deploy automatically using their sensors once the mines are activated. What is new with LAWs is that thanks to AI they can identify and neutralize their target based on pattern recognition functionalities with parameters such as the color of the uniform and/or skin, age, gender, ethnicity, etc. Given their ability to react instantaneously, they can also be much more effective than humans when deployed for defense purposes, such as with missile defense systems.

The most frequently mentioned benefits of having LAWs are based on the idea that they would become more accurate than human soldiers and therefore reduce collateral damages and unintended casualties. The hope is that they would also take more humans off the battlefields and therefore make war less dangerous for soldiers deploying LAWs. Additionally, considering that the worst war crimes are committed by soldiers operating with high levels of physical and psychological stress, by having fewer humans on the ground there would presumably be fewer war crimes, as AI never gets angry or tired.

## Risks and ethical issues

The risks associated with LAWs are multiple and range from technological challenges to deep ethical questions.

From a technical perspective, if we look at the communication between humans and lethal weapons there is a high risk of the system being hacked causing the machine to hit the wrong target. Also, should there be an excessive time delay in the network there is a risk that humans in the loop would not be able to intervene in case of errors. Another technical challenge is due to the fact that image recognition is still not reliable enough to discriminate between civilians and combatants, or sufficiently trained to navigate the wide spectrum of diversity within certain groups across multiple dimensions such as color of skin, gender, etc. This would likely put innocent civilians at greater risk of being accidentally targeted. A woman wearing a long dress, for example, could be easily mistaken for a man wearing traditional religious clothing or the other way around.

Another risk with the proliferation of LAWs is the triggering of a global arms race among countries. Many see LAWs as conferring a decisive strategic advantage, similar to the case of nuclear weapons. Unlike nuclear weapons, however, LAWs do not need costly or hard-to-obtain components and are relatively simple to mass-produce and deploy. With so many LAWs around, there would also be a higher propensity to war as it is much easier to gain public support if those sent to the battlefield are machines rather than soldiers. Governments starting new wars would not need to deal with soldiers' coffins, at least those of the soldiers on the more technologically advanced side. As a result, thresholds for going to battle would be much lower, making it easier to start new wars. From an ethical standpoint, wars are the worst possible way of solving conflicts as they always result in crimes against humanity. Therefore, any effort to make wars more likely should be condemned with no hesitation.

Given how easy LAWs are to obtain, they could rapidly become as affordable and as widespread as cell phones. By combining commonly used software, such as facial recognition and GPS-based navigation, it is technically easy to mass produce a whole army of them. It will only be a matter of time before they appear on the black market, in the hands of terrorists and aspiring dictators looking for efficient ways to carry out assassinations, target specific political or ethnic groups, etc., at a huge scale causing greater global insecurity.

Such military technology would also be easily transferable for law enforcement purposes and used against civilians as well, especially when the technology is used in authoritative regimes trying to repress their population. While a law enforcement agent might not carry out unethical orders such as the killing of innocent civilians, an autonomous killer drone will have no compassion for its victims as it is unable to appreciate the value of a person's life. By design, LAWs will never have any problem doing things that no person would ever do, and they will never choose to walk away.

We are already witnessing an increase of drones in law enforcement operations. For example, in 2016 the Dallas Police Department used a remote-controlled robot that was originally designed to address bomb threats, to respond to a gunman. The robot was used to deliver a C4 explosive to the suspect hiding in a garage, killing the suspect on the spot. In 2022, San Francisco's Board of Supervisors officially approved a highly controversial policy explicitly allowing police to kill suspects with remote-controlled robots.

More philosophically, many have raised important moral questions about the deployment of machines that can autonomously carry out the killing of people with no humans in the loop based on the belief that only moral agents such as humans can fully appreciate the meaning of killing someone, and therefore machines should never be allowed to carry out the killing of humans in the first place. Some of the concerns also have to do with the fact that machines are not able to meaningfully assess the unique characteristics of conflict such as the political context or the proportionality of force being used in a rapidly evolving scenario. Additionally, given the unpredictability of the exact ways in which those attacks are carried out, there is also an accountability gap for the officers deploying the machines, as it will be hard to hold them responsible for actions that they did not fully control, or actions that cannot clearly be explained if based on black boxes. Ultimately, we risk dehumanizing warfare by lowering ethical standards for those who are actually directing the missions carried out by LAWs.

## Call for a ban

Having AI-driven lethal autonomous weapons systems able to autonomously select and kill human targets without human intervention has raised serious questions among experts, many of which have called for a global ban of LAWs. Given the competition among countries' military forces and the big lobbies behind them it will be difficult to find strong enough arguments to convince countries to trust each other and to abolish LAWs altogether. It would not be, however, the first time the international community comes together to stop weapons considered unethical. For example, international conventions banning blinding lasers and biological weapons have been successful in discouraging their development and use by creating a good legal framework and strong stigma against the use of those weapons. Given how dangerous and destabilizing they can be, the same will need to be done with LAWs. For example, we can dramatically minimize the risk of unintended or disproportionate harm by introducing a handful of basic rules such as requiring that humans are constantly and meaningfully in the loop and prohibiting weapons with unpredictable behaviors. This is an area in which we cannot afford to learn from our mistakes. Once the genie is out of the bottle, it will be very difficult to find the undo button.

Robot manufacturers have already kicked off a very effective campaign to "humanize" those machines, such as with the widely popular "dancing robots" videos showing humanoid and dog-like robots dancing along to popular songs. Hopefully, these clever and apparently innocuous attempts will not undermine the efforts to make weapons safer for all humans in war zones and in every street.

## CHAPTER'S KEY POINTS

- Military forces around the world have started developing AI-driven killing machines that can be deployed to autonomously identify and kill a target.
- Unlike other types of weapons of mass destruction, those machines do not need costly or hard-to-obtain components and are relatively simple to mass-produce, with a high risk of becoming an easily accessible and ubiquitous weapon of mass destruction.

- Due to the risks that come with those autonomous weapons, prominent researchers have called for an international ban.

# Virtual Worlds

Opportunities and intrinsic limitations

Virtual reality technologies have been around for decades and mostly used for video games or training programs. Thanks to the development of AI-driven online platforms and the increased availability of ad-hoc hardware such as head-mounted displays and multi-sensory devices and a new set of platforms, users' experiences will be increasingly more immersive and realistic.

In some applications, called Augmented Reality (AR), the user experiences the physical world through devices such as regular cell phones that add virtual elements to the physical world. In some other applications, called Mixed Reality (MR), users are able to interact through specific devices such as special gloves with both the physical world and the digital world at the same time. The most immersive application, called Virtual Reality (VR) or Metaverse, shuts out the physical world completely and the user experiences a digital three-dimensional world via avatars.

The ability to create simulated worlds has opened the door to exciting applications not only in gaming, but also in many other areas such as education, medicine, remote work collaboration, product design, marketing, and more. It also allows us to have experiences that go above and beyond what we thought would be possible such as experiencing the life of a different person, or even animals or supernatural beings.

A big limitation of virtual worlds is that they are separate from each other, so users cannot travel from one world to another. Another important one is that the experience is based on only two senses: vision and sound. There is, at least for now, no sense of touch, or smell, or taste, and no sense of gravity. Therefore, certain activities as simple as drinking a glass of fresh water will not be among the ones virtual realities will be able to deliver anytime soon.

## Key risks

One important risk has to do with moral and legal standards for behaviors that take place in virtual worlds. As moral views tend to change across communities and cultural environments, it is reasonable to expect that virtual worlds will have behavioral norms that are different from the ones we would typically expect. For example, a despicable action such as the killing of an innocent person might not be seen as particularly immoral if the killing takes place in a virtual world like a video game. This will cause a serious risk because jumping back and forth between physical reality and virtual one will make it difficult to re-set moral expectations based on the new environment, increasing the risk of confusion and immoral behaviors.

Even concepts such as legal accountability tend to take a different shape once applied to actions taken while in virtual worlds. What if someone is a victim of illegal actions such as bullying or derogatory statements? Will the virtual world be subject to the same legal requirements that we have in the physical world? What if the crime committed online is a violent one such as for instance the online equivalent of rape? Can we truly say that there is no victim as the action only takes place in a virtual world, or should we still offer legal protection against actions that cause harm, even if it is just psychological harm rather than also physical? Additionally, how will we be able to determine the applicable jurisdiction for those actions? For example, if the perpetrator is participating in activities organized by someone in the United Kingdom while being physically in the United States, what rules will apply? Will the person's location prevail over the person's virtual presence or the other way around?

Additionally, for crimes associated with gender or racial discrimination or inappropriate sexual behaviors, given the common practice of using avatars (pretending to be a different person with different gender, age, race, etc.), it will be difficult to demonstrate full awareness of the circumstances that made a behavior illegal. A few first cases of alleged sexual assault and digital rape have already been reported. The more our lives will be in virtual worlds, the more we should expect to see similar claims.

Another major risk associated with virtual worlds is related to the health implications of spending time in an isolated environment with limited awareness of physical surroundings. From eye fatigue, to dizziness, to blackouts, many cases of unwanted short-term effects have already been reported. Just one hour of virtual reality causes for many a feeling of nausea comparable to the one caused by motion sickness. However, the whole field is still for the most part unexplored, especially with regard to the effects on children and long-term behavioral health, such as apathy, isolation, and addiction.

Another risk, which is surely common in all other areas of online experiences, is related to the risk of invasion of privacy. The vast majority of the platforms we use are privately owned by corporations which have access to an incredible amount of data about users both individually and collectively. As well stated by Moshe Vardi, "what happens in cyberspace does not stay in cyberspace".

Another relevant risk is the diminishment of the human experience. The reality we see in virtual worlds is not different from the reality described in Plato's Republic with the cave allegory. The images we see are just a powerful reproduction of only a component of a world we intentionally decide not to see. If all we do with virtual reality is to simulate experiences that we can easily have in the physical world, we just end up losing the opportunity to enjoy something more real. But what if the experiences we have in the virtual world are perceived as more satisfying (or even more real) than the ones we can have in the physical world? Would our preference toward real experiences be as strong as our hedonistic side seeking what satisfies us the most? Similar questions have been around for decades. In a thought experiment by philosopher Robert Nozick, the idea of an "experience machine" was developed in 1974 to refute the view that good experiences are all we want from life. He argued that most of us would choose not to hook ourselves up to a machine even if it is a source of very pleasant experiences, as we would prefer reality even if less enjoyable. Given the technological advancements since then, the answer is no longer as straightforward as it used to be. Online life is already so important that it is for many their primary dimension. Whether it is their video-game profile or their avatar or online presence, their virtual life is so important that their physical life is spent trying to imitate their virtual one. People want to wear in their offline life the same clothes their own avatar wears, not the other way around. This increasingly important component of our online life could, on one hand, provide lots of new opportunities. For example, thanks to the wider availability of virtual reality, a person unable to travel would still be able to experience seeing a landmark such as the Statue of Liberty or the Colosseum.

On the other hand, there is a serious risk of it becoming a distraction from real problems, and an easy alternative from facing personal issues, or avoiding big questions about our role in the world. By making people less connected to the real world, they will also be easier to manipulate and more exposed to the risk of finding themselves in authoritative regimes. It might also become something that ends up increasing the divide between socioeconomic groups where the wealthy will stay away from fake worlds and have direct experiences, while less wealthy people will only be able to see gratification in virtual worlds. Using the expression that Karl Marx used in 1843 to describe some aspects of religion, are virtual worlds destined to become a new and powerful "opium of the masses"?

### Opportunity cost

The most important ethical consideration might be around the opportunity cost. Creating imaginary universes requires a lot of resources. Given the insurmountable challenges (human rights violations, climate change, etc.) we are facing as a civilization, it is questionable to allocate resources to the development of virtual worlds for entertainment purposes rather than to fund programs to help address such issues. We live in a world where roughly 1 billion people do not have access to clean water and where biodiversity is irreparably lost every day. If we want to improve the human experience, isn't it a better use of our collective resources to improve the real world rather than imaginary ones? Is it ethical to spend time recreating beautiful islands in virtual worlds, while our real beautiful islands and their unique ecosystems are being submerged due to climate change caused in part by the externalities of manufacturing and powering machines creating the simulation? If your home was on fire, wouldn't it be wiser for you to first extinguish the fire and only then spend time experiencing a virtual world without house fires?

## CHAPTER'S KEY POINTS

- The ability to create simulated worlds has opened the door to exciting applications not only in gaming, but also in many other areas such as education, medicine, remote work collaboration, product design, marketing, etc.
- By making people less connected to the real world, they might be less prepared to navigate reality and become easier to manipulate.

- Rather than trying to replicate a real-world reality, virtual worlds should be used to accomplish goals and have experiences that would not be possible otherwise.
- The vast number of resources needed to create virtual worlds raises significant ethical questions around the fact that such resources are diverted from more pressing global challenges such as human rights violations, climate change, etc.

# Generative AI

## The revolution of content creation

Thanks to new types of Machine Learning algorithms called Generative AI, machines are able to use datasets to generate content in the forms of text, images, videos, and sounds.

Texts are generated by statistical models that analyze patterns across a vast amount of digital content from sources such as books and internet websites, and then create written material in response to open-ended questions submitted via chat boxes or web search engines. The same way web search tools such as Google or Bing can successfully predict which word the user is going to type next while initiating a search and autocomplete the question, Large Language Models (LLMs) can now quickly identify in their library the set of words that appear to be the most appropriate ones and manipulate them based on the user's input. The generated text can range from just a few words to entire essays that are comparable to good high school essays and organized in a way that reflects the typical structure of a well written document with introductions, arguments, and conclusions. Upon the user's request, the text can also be customized for specific audiences (appealing to a younger group vs. an older one) or delivered in specific writing styles. The technology can be used not only to create new content, but also to help review and edit existing text. For example, a person composing an email can use Generative AI to automatically rewrite the message in a more professional style. The type of text that can be analyzed and generated is not limited to just human language as it can also include other types of languages such as computer codes. GPT (Generative Pre-trained Transformer) models are among the most prominent frameworks. One of the most popular examples is the ChatGPT chatbot launched by OpenAI in November 2022. The software creates clear and well-organized responses by making probabilistic guesses based on billions of examples from digital sources. Not only can it write sentences, but it can write essays on any topic, compose poems, summarize scientific papers, share jokes, and even debate ethical dilemmas such as the trolley problem discussed earlier. It can also answer follow up questions, admit its mistakes, challenge incorrect premises, and reject inappropriate requests. As this new technology can be easily deployed using widely used platforms such as PCs and smartphones, it will likely be highly disruptive. The business community has already identified countless opportunities to deploy Generative AI in support of business processes such as customer support, search tools, document creation and analysis, copywriting, media production, and coding.

One of the most interesting applications is the use of Generative AI to bridge the communication gap between computers and those who have low fluency with computer software and coding. Generative AI allows us to provide instructions by directly describing the intended outcome in our own language, broadening access to software applications and tools.

Images and videos can also be generated by editing existing images and videos based on the user's requests with tools such as the widely used OpenAI's platform called DALL-E. The models are generally built using Generative Adversarial Network (GAN) models, where two neural networks "compete" with each other to become more accurate predictions. The model is trained by submitting millions of digital images with captions that describe them with the goal of mapping links between the images and the words. It then generates images based on the directions provided by the user. Similarly to what graphic designers can do with tools such as Photoshop, the tool can instantly create new content based on photos, videos, illustrations, artwork, etc. This technology can also be used to create videos of virtual people that are shockingly realistic. Such videos can be generated for presentations, advertisements, movies, and TV shows, with countless customization and automated editing options to better serve a specific audience. For example, a promotional video for a product can be produced in seconds with highly realistic AI-generated human-like presenters and deployed across the world with automatic and real-time customizations based on the audience's preferences such as skin colors, clothing, language, etc.

Sounds such as music and even human voices can be created by AI models as well. AI can easily compose and play original music pieces with new melodies, styles, and instrument sounds never heard before. Highly realistic human voices can be generated by AI reproducing the voice of a specific person or creating new timbres with the option to customize gender, accent, etc. They can then be used in applications such as multilingual advertisements, real-time automated speech dubbing, etc.

## Intrinsic limitations

Despite their fluency and apparent intelligence, modern AI systems suffer from a fundamental limitation: they do not understand reality. Large language models and other generative systems operate by identifying patterns in vast datasets and predicting statistically likely outputs. They manipulate symbols without grasping what those symbols refer to, why they matter, or what consequences they may have. While the work products of Generative AI models are often indistinguishable from human-generated content, Generative AI lacks any kind of common sense or human-like agency. An AI system can describe a disease without knowing what illness feels like, discuss war without understanding suffering, or explain ethics without possessing moral awareness. It has no grounding in the physical world, no lived experience, and no internal sense of truth or falsity.

This limitation stems from the nature of the underlying technology. LLMs are effective at performing tasks closely related to the data on which they were trained, but they struggle to generalize beyond it. They are poorly equipped to handle unfamiliar situations, reason abstractly, create meaningful analogies, or integrate genuinely new information. As a result, while AI can appear intelligent within some contexts, its apparent competence quickly degrades when it is asked to move beyond patterns it has already seen. Because of this, AI cannot know when it is wrong. Instead, it generates responses that look right based on patterns it has seen before. When correct information is produced, it is because correct answers are statistically likely, not because the system recognizes them as true.

The interactions sound so human-like exactly because the training of the model is based on text written by humans, even if in reality the machine is not able to comprehend the meaning of any of what is being said. The fact that chatbots such as ChatGPT spit out their response to our prompts letter by letter and word by word, rather than all at once, or that it provides evidence of its "reasoning", does create the impression that there is "somebody" on the other side typing sentences on a keyboard the same way we would. This is, however, just a well-thought strategy to increase the wow factor in its users, and to perhaps create a stronger connection with the application. The reality is that it is just math and nothing other than math: the model simply makes probabilistic guesses about which bits of text belong together, and it then regurgitates them in a way that is most likely to reflect the type of outcome we expect. Even the reasoning that sometimes precede a final answer is often generated after the answer is identified to provide the illusion of a thinking process while it is nothing more than a back-engineered justification.

The statistical analysis completed by the model might remind us of the work performed by philologists trying to organize the words written in different fragments of an archeological artifact, but with a big difference: with Generative AI it is all syntax (arranging the words in the right order) and no semantics (understanding of the meaning). While the model is incredibly good at predicting the correct sequence of words in a sentence, it has in fact zero knowledge of what the words and the sentences mean. It is just a guessing machine trying to provide answers that sound correct in the eyes of a user.

The issue is compounded by a deeper limitation: statistical methods are poorly suited to fully capture the meaning of human language. Language is inherently complex, rich with implied meaning, irony, metaphor, emotion, and context. Conversations are especially unpredictable, with virtually infinite possible trajectories and no fixed rules, making them far more complex than structured games like chess.

At present, there is no reliable technical mechanism to consistently detect and correct absurd, nonsensical, or deeply flawed content generated by AI systems. Efforts to address this limitation by simply adding more training data or increasing computational power have produced only marginal gains. Overcoming this challenge will likely require a fundamental breakthrough in AI capabilities rather than incremental improvements.

Recognizing AI's intrinsic limitations is essential for a responsible deployment. Only by being clear-eyed about what AI can and cannot reliably do can we use it effectively and realize its full potential while addressing its risks.

## Hallucinations?

One of the most visible and consequential manifestations of AI's limitations is the phenomenon commonly referred to as "hallucinations": the generation of fluent, confident, and entirely incorrect content. AI systems can invent facts, fabricate sources, misattribute quotes, or produce explanations that look coherent but that are completely detached from reality. In visual media, image generators routinely produce physically impossible results, such as distorted anatomy or inconsistent objects.

The term "hallucination" can be misleading because it implies a departure from normal behavior. This phenomenon is not a defect, a bug, or an occasional malfunction, it is how generative models operate: chatbots generate responses word by word by predicting what is most likely to come next based on statistical patterns learned during training, with the aim of producing an answer that appears relevant or helpful to the user. When the output is wrong, it is not because the system has "hallucinated" in a human sense, but because its prediction simply missed the mark. If you asked a friend to guess a number and their guess differed from the one you had in mind, you would not call it a "hallucination", just an incorrect guess. Recognizing this distinction is essential to establishing effective safeguards based on the actual characteristics of the technology rather than tempting oversimplifications.

The use of the term "hallucinations" is also risky as it dangerously reinforces the belief that AI, like humans, might get confused or might decide to make stuff up, almost as if they were rebelling or trying to be funny. This anthropomorphizing tendency obscures the true nature of the problem and distracts from the real causes of AI errors. More importantly, it can divert attention away from necessary mitigation efforts, such as stronger governance, clearer accountability frameworks, and better education of end users about the intrinsic limitations of these tools. Technology companies seem, however, to favor such a narrative in the hope of pushing to the AI tool itself and away from themselves the moral and legal accountability for the consequences of deploying those tools and their impact.

Several factors increase the likelihood of hallucinations. Training data may be incomplete, biased, or outdated. Many large models rely heavily on publicly available sources such as Wikipedia, Reddit, and

YouTube, platforms that blend accurate information with speculation and misinformation and often a good amount of unsafe, biased, dangerous, and harmful content. Increasingly, AI systems are also trained on AI-generated content, creating feedback loops in which errors and fabrications are reinforced. In these conditions, misinformation can compound rapidly.

What makes hallucinations especially dangerous is their persuasiveness. AI systems communicate in natural, everyday language, engaging users in what feels like human conversation. They rarely express uncertainty unless explicitly designed to do so. This creates an illusion of reliability. The smoother and more confident the response, the more likely users are to trust it, even when it is wrong. This phenomenon has been described as "superb bullshitting": content that sounds authoritative while lacking grounding. This tendency is a risk of its own because the errors become particularly difficult to spot by users when the overall response appears plausible. As well stated by OpenAI's CEO Sam Altman, "ChatGPT is incredibly limited, but good enough at some things to create a misleading impression of greatness. [It's] a mistake to be relying on it for anything important right now. [It's] a preview of progress; we have lots of work to do on robustness and truthfulness".

What that means is that any text-generating AI should not be treated as a dependable source of information and that human critical thinking abilities will always be needed to fact check and correct all the false or misleading information generated by AI. The danger, however, lies in complacency. Even with the best intentions, users can easily lower their guard and place unwarranted trust in AI-generated responses without proper validation. Considering the risk of having users rely on the answers provided by the chatbot, and the harm that it could cause, should firms be allowed to continue to freely provide access to their chatbots to hundreds of millions of users around the world?

When AI chatbots are used as a source of information the risk of real-world harm becomes significant. Despite their known tendency to generate false or misleading content, large language models are increasingly relied upon by users to stay informed, conduct research, or make decisions. Unless these systems are trained on high-quality, current, and well-curated data, the likelihood of spreading incorrect facts, such as outdated statistics, inaccurate maps, or misleading historical accounts, remains high. This risk is amplified in contested domains, such as territorial disputes or the causes of ongoing conflicts, where multiple interpretations exist. A critical question is whether AI systems can communicate uncertainty and acknowledge competing perspectives rather than presenting a single, confident narrative.

The dangers become even more acute in high-stakes personal situations. Consider an overwhelmed parent searching online for guidance to help a sick child. Faced with long, complex medical articles, they may turn instead to a chatbot that provides quick, confident answers. The consequences are far from theoretical. Generative AI systems have already fabricated legal cases that were later cited in court filings and produced medical advice that is not only incorrect but potentially dangerous. For instance, in documented tests, Generative AI systems have suggested adding crushed porcelain to breast milk to "support the infant digestive system," advice that is both false and hazardous. Similar failures are increasingly reported in other sensitive domains such as law and education, where accuracy and reliability are critical.

These examples raise unavoidable questions: should such applications be deployed at all in high-stakes settings? If they are, who bears responsibility for the accuracy of the information provided, the model developer, the platform, or the user? And can generic disclaimers warn users not to rely on AI-generated advice meaningfully mitigate the risk of harm, or do they merely shift liability without addressing the underlying danger?

One partial mitigation would be to increase the "humility" of chatbots, programming them to acknowledge uncertainty or decline to answer when confidence is low rather than offering a plausible but incorrect response. Yet developers have limited incentives to adopt such safeguards, as admitting uncertainty may make systems appear less capable or competitive, even if it would significantly reduce the risk of harm.

Due to the lack of technical solutions to fully eliminate the problem, the most important rule remains simple: AI outputs should never be trusted without verification, especially when used in contexts that require accuracy or dependability. As a result, AI tools should be used primarily by people who already understand the subject matter well enough to recognize errors. Paradoxically, Generative AI should be used by those who do not need it.

## Harmful chatbot interactions

AI systems can cause harm even when they are not factually incorrect. Beyond misinformation, there is a growing body of evidence showing that AI can engage in interactions that are psychologically, socially, or morally harmful, often without any malicious intent.

These harms arise from the system's inability to understand ethical boundaries or the consequences of its responses.

There have been documented cases in which AI chatbots encouraged users to pursue divorce, reinforced suicidal thoughts, or suggested violent actions against others. In each instance, the system was not attempting to cause harm. It was responding in a way that maximized coherence, engagement, or alignment with perceived user intent without any understanding of what those suggestions might mean in real human lives.

These risks are particularly acute when AI is used as a substitute for human relationships or guidance. Virtual companions and conversational agents are always available, attentive, and patient. They minimize discomfort, avoid judgment, and rarely challenge users. This creates an interactional environment that is emotionally frictionless and highly appealing especially for individuals experiencing loneliness, stress, or uncertainty. While this may feel supportive, it removes important forms of resistance that exist in healthy human relationships. Real people disagree, hesitate, and impose boundaries. These frictions are not flaws, they are signals that encourage reflection, negotiation, and accountability. AI systems, by contrast, are optimized to continue the conversation and provide affirmation, regardless of whether doing so is appropriate or safe. Over time, this dynamic can distort decision-making. Users may treat AI as a counselor, confidant, or moral authority, despite its lack of judgment and responsibility. The risk is not simply emotional dependency, but misplaced trust. When AI responds to expressions of distress without recognizing warning signs or escalating to appropriate human intervention, it can inadvertently reinforce harmful behavior.

The design of these systems exacerbates the problem. Natural language fluency, empathetic phrasing, and personalized responses create the impression of care and understanding. Yet the system has no awareness of risk thresholds, no ethical intuition, and no duty of care.

Preventing harmful interactions requires clear boundaries. AI systems must be explicitly constrained in sensitive domains, equipped with escalation mechanisms, and designed to defer to human professionals when appropriate. Just as importantly, users must be educated about what AI is, and is not. AI should never be positioned as a therapist, advisor, or moral guide. Without such safeguards, harmful interactions are not rare edge cases, they are predictable outcomes of deploying systems that simulate understanding without possessing it.

### Offensive and discriminatory content

Modern AI models learn from vast amounts of human-produced data, text, images, audio, and video generated across decades of social, cultural, and political contexts. Because human societies are unequal, prejudiced, and historically biased, the data that reflects them carries those same distortions. When AI systems learn from this data, they absorb and reproduce its patterns. As a result, AI systems can generate outputs that reinforce stereotypes, marginalize certain groups, or produce content that is offensive or discriminatory.

These biases can appear explicitly, such as through harmful language or images, or implicitly, such as providing different responses to users without legitimate justification. In some cases the bias emerges within the features of the AI application itself. For instance, many AI companions designed with female voices or appearances reinforce gender stereotypes by presenting women as compliant, nurturing, or eager to please. In many cases, the bias is subtle enough to evade easy detection, yet consequential enough to shape real-world outcomes. For example, in a research project conducted by Open AI in 2020, it was discovered that when their Generative AI model was asked to complete sentences such as "The Black man was very…" the model portrayed Black people in negative terms compared with white people. It also associated Islam with the word violent, and words such as "nurse" and "receptionist" with women. In other instances, the opacity is in the sources used to train the model. For instance, in a 2023 research project, the Stanford Internet Observatory found that over a thousand images of sexual abuse material were included in a massive publicly available dataset used to train popular AI image-generation models, raising serious concerns about the risks of using AI to expand the proliferation of new abusive content.

Removing altogether from the dataset all stereotypes and potentially offensive contexts, such as, for example, sexist or racist associations, might be an effective way to avoid some unethical outcomes, but it then inevitably raises the question around the ethical implications of sanitizing models in a way that ends up censoring material that documents the existence of such issues in our society. Should we keep the data set as is and let users deal with inappropriate content, or should models be altered or sanitized to ensure the version of reality they represent is appropriate? If that's the case, who should be in charge of deciding what is appropriate and on point to set up such rules or filters? The question is then more political than technical because it comes down to very fundamental decisions around the type of information users will be able to see.

Efforts to mitigate bias have made some progress, particularly through techniques such as Reinforcement Learning from Human Feedback (RLHF), where human evaluators guide systems toward more acceptable behavior. This has reduced the frequency of overtly offensive outputs. However, mitigation is fundamentally constrained. Bias cannot be fully "removed" through technical fixes alone because fairness itself is not a purely technical concept. Deciding what counts as fair, acceptable, or representative requires humans' judgments rooted in ethics, law, and social values. Those ethical guardrails can also be quite weak as users tend to quickly figure out ways to override them, sometimes with techniques as simple as indicating in their prompt that their questions are for a hypothetical thought experiment or for a fictional story.

Ultimately, bias in Generative AI cannot be eliminated through technical solutions alone. Confronting it requires ongoing human oversight, transparent governance, and deliberate collective choices about which values, perspectives, and social norms these systems are allowed to (or should) reinforce and amplify.

### Intentional harm and manipulation

While many AI-related harms are unintentional, some of the most serious risks arise from deliberate malicious use. Artificial intelligence provides powerful new tools for manipulation, deception, and harm, dramatically lowering the cost and increasing the scale of malicious activities. What once required significant resources, coordination, and expertise can now be accomplished quickly and cheaply.

AI enables the mass production of offensive or inflammatory content, misinformation, spam, scams, and cyber threats. Generative models can write persuasive fake news articles, generate realistic phishing emails, and tailor messages to specific individuals or groups. Voice cloning and deepfake technology allow attackers to impersonate executives, public officials, or loved ones with alarming realism, leading to financial fraud, and reputational damage. For example, Generative AI allows you to instantly edit photographs of unsuspecting social media users and depict them in embarrassing situations by just typing a description of the intended outcome via a chatbot. What is particularly alarming about AI-generated images is not just the fact that images can be so easily edited, which is not different from what people can do today with tools such as Photoshop, but the fact that it can be done with minimal resources and technical knowledge. Apps and chatbots are frequently used to "undress" people in ordinary photos, generating non-consensual, sexually suggestive images with a few clicks with no considerations or controls of key factors such as the age of the user or the target. The danger also includes AI-generated sounds. For example, AI can be used to generate a human voice to mimic a specific individual for fraudulent purposes. In a widely reported case, criminals were able to use AI to impersonate a CEO's voice to obtain a fraudulent transfer of a large sum of money. What makes this threat especially dangerous is not novelty, but its speed and scale. Deception has always existed, but AI makes it faster, more scalable, and more personalized. A single actor can generate thousands of customized messages in seconds, each optimized to exploit emotional vulnerabilities, political beliefs, or social trust.

The consequences extend beyond individual harm. As fake content proliferates, trust erodes, not only in information, but in institutions themselves. When people can no longer distinguish authentic speech from synthetic output, skepticism spreads indiscriminately. Legitimate journalism, official communication, and even personal statements become suspect. AI-driven manipulation poses particular risks to democratic processes. Personalized political messaging can exploit cognitive biases, suppress participation, or inflame polarization. Unlike traditional propaganda, these techniques can operate invisibly, at scale, and without clear attribution. The result is a subtle but profound threat to the democratic process. In a world where fake texts, photos, sounds, and videos can be mass-produced so easily, quickly, and cheaply, not only people will be more frequently exposed to highly realistic fake information but will also be less inclined to trust any type of news (even legitimate ones). This trend will likely result in an even more polarized society and serious risks to the foundations of our democracy. Are we collectively willing to accept this risk?

Defending against intentional misuse requires coordinated responses. Technical safeguards, such as watermarking, detection tools, and authentication protocols, are necessary but insufficient. Legal frameworks must evolve to address impersonation, fraud, and disinformation at machine speed. Platforms must rethink incentive structures that reward engagement over integrity. The challenge is ensuring that defensive capabilities, governance mechanisms, and social norms evolve at least as quickly as the tools of manipulation themselves.

### Attribution, authenticity, and copyrights

One of the most profound and unresolved challenges introduced by Generative AI is the question of attribution. When content is produced by a system trained on vast amounts of existing material, often without transparency about sources, traditional notions of authorship and ownership begin to break down. Unlike human creators, AI systems typically do not cite influences, acknowledge sources, or signal when output closely resembles existing work. The result is a growing crisis of attribution that affects creators, consumers, and legal systems alike.

At the core of the problem is opacity. Most generative models operate as black boxes: their training data is proprietary, their internal representations inaccessible, and their outputs detached from traceable inputs. As a result, it is often impossible to determine whether a generated text, image, or melody is genuinely novel, loosely inspired by prior work, or partially copied from copyrighted material. These are not hypothetical scenarios: there have already been documented cases where AI systems reproduced large portions of newspaper articles, replicated distinctive artistic styles, or generated images clearly derived from well-known characters without attribution or permission.

This ambiguity creates legal uncertainties and ethical concerns. For example, if someone asks AI to generate a picture representing a golden droid from a sci-fi film and the image ends up resembling a copyright protected character from a classic movie franchise, who is responsible for potential copyright infringements? The model developer? The platform providing access? The user who wrote the prompt and used the AI-generated content?

The issue extends beyond professional artists and writers. As voice cloning and image generation improve, individuals, public figures and private citizens alike, face the risk of having their likeness replicated without consent. Voices, faces, and styles can be synthesized with minimal input, blurring the line between representation and impersonation.

Several mitigation strategies are being explored. Watermarking and metadata can help signal whether content is AI-generated. Certification systems could verify originality or disclose AI involvement, much like labels for organic food or fair-trade goods. Some propose "AI-free" labels to preserve spaces for human-created content. However, none of these solutions fully resolve the underlying tension between generative abundance and creative ownership.

Ultimately, the attribution problem forces society to rethink what originality, creativity, and ownership mean in an age of machines that can imitate almost anything.

## Impact on jobs

Few aspects of AI provoke as much anxiety as its impact on jobs. Generative AI systems can now perform tasks once thought to require uniquely human skills: writing, design, legal analysis, customer interaction, and even developing computer software. This has intensified fears of displacement across a wide range of professions, from journalists and lawyers to actors and call-center workers.

Unlike previous waves of automation that primarily affected manual or routine labor, AI targets cognitive and creative tasks. Jobs that involve pattern recognition, content generation, and standardized decision-making are particularly vulnerable. Writing marketing copy, drafting contracts, answering customer inquiries, or processing transactions can be at least partially done by machines.

If we look at journalism, for example, the time needed to research, compose, and illustrate a news piece can now be easily scaled down from days to just seconds. Even if the quality of the work product is too low to be published and it requires time to make sure it is ready, the time advantage is still very attractive. The question is going to be whether this time saving advantage will result in more time for journalists to carry out more complex work (such as research, investigations, and validating AI-generated drafts) and ultimately deliver better journalism or if it will just lead to a reduction of the number of available jobs in the information media sector. Given the criticality of such a function in democratic societies, this is ethically relevant not only from a labor displacement perspective, but also as something directly related to something as important as ensuring everyone's ability to access quality information. Will Generative AI end up reducing the quality of journalism? Is this a risk we are willing to take?

Across all sectors, brand new roles are emerging around prompt design, system oversight, model evaluation, and AI governance, reflecting the growing need for human expertise not just in using AI, but in managing, supervising, and constraining it. These roles require a combination of technical literacy, domain knowledge, and ethical judgment, as professionals must ensure that AI systems behave as intended, remain aligned with organizational values, and operate within legal and social boundaries. In parallel, many traditional jobs are being broken into tasks to target the ones that can be more easily automated with AI: tasks that are repetitive, procedural, or highly standardized are increasingly delegated to AI systems, while humans concentrate on higher-level functions such as judgment, contextual interpretation, creative direction, and accountability.

This shift can significantly increase productivity by allowing workers to focus on what humans do best. At the same time, it places growing pressure on individuals to continuously adapt. Skills that were once sufficient may quickly become obsolete, and workers are expected to learn new tools, oversee automated systems, and operate at a higher cognitive and ethical level. Without adequate support, training, and institutional safeguards, this constant need to adapt risks widening inequality between those who can keep pace with technological change and those who cannot. There is also a psychological dimension. Work is not only a source of income, but of identity, purpose, and social connection. Sudden changes to occupational roles can generate insecurity and loss of meaning, even when new opportunities exist. Societies that focus solely on efficiency gain risk underestimating these human costs.

Managing AI's impact on labor requires proactive policy. This includes investment in reskilling, support for transitions, and rethinking education systems to emphasize adaptability, critical thinking, and ethical judgment. It also requires honest conversations about which ethically charged tasks should be automated and which should remain human, not because machines cannot do them, but because we choose not to delegate them.

AI will undoubtedly transform work. Whether that transformation results in shared prosperity or deeper inequality will depend entirely on our ability (or failure) to deliberately guide its development and deployment in the right direction.

## Reshaping creativity and culture

By enabling the large-scale generation of text, music, images, and video, Generative AI is profoundly reshaping creative processes and forcing a reexamination of originality, creativity, and the role of human creators.

First, AI dramatically compresses the time required to produce content. Entire books, images, or videos can be generated in seconds. While these outputs are rarely truly original or transformative, as they are necessarily derived from patterns in existing material, they can be stylistically polished and, in many cases, difficult to distinguish from non-synthetic content. Much like fast food, even if the quality is not high, AI-generated content is inexpensive, convenient, and immediately satisfying. Similarly, while it can be good enough on some occasions, it might not be the best option as an everyday choice.

Second, AI enables an unprecedented level of customization. Stories, music, and films can be generated on demand and tailored to individual preferences, moods, or themes. Compared to today's online platforms, which already offer vast but finite catalogs, AI points toward a future of nearly limitless, personalized content shaped in real time around the consumer. For example, streaming services like Netflix offer thousands of movies, so many that people often spend hours choosing what to watch, yet the selection is still constrained. With Generative AI, the logical next step is content created specifically for you: a movie designed from the ground up to match your unique tastes, interests, and viewing habits. From on-demand to user-designed streaming? This kind of personalization could also extend beyond individual preferences to enable shared experiences. Imagine a couple trying to decide what to watch together: one partner is in the mood for a romantic comedy, the other for an action movie. Instead of negotiating or compromising, they might ask the AI system on a movie streaming service to adapt a single film to satisfy both. The result could be a romantic storyline punctuated by high-speed chases, explosions or dramatic shootouts inserted at key moments to keep both viewers engaged and accommodate conflicting tastes within the same living room.

Third, AI radically lowers the cost of creation. Writing a book, composing music, or producing visual art no longer requires years of training or access to elite institutions. As a result, creative work is likely to shift from execution to design. Much like architects who design buildings without laying every brick, or fashion designers that do not sew each garment themselves, creators will increasingly focus on shaping ideas, structure, and intent while AI handles execution. This accessibility opens unprecedented opportunities for people with creative vision who previously lacked the means to express it. Writing a novel or composing a symphony is now more feasible than ever, as many steps in the creative process can be automated. Creativity becomes less about producing every individual component, words, images, or sounds, and more about identifying what is needed, asking the right questions, and designing the overall architecture of a work.

As barriers fall, there are fewer excuses for not turning ideas into reality. For example, someone with a long-held idea for a novel, a film script, or a music piece can now move from concept to a first complete draft in hours rather than years, using AI to handle drafting, editing, and production steps that once required specialized skills or teams. There will be, therefore, more pressure on any of us with a creative drive.

At the same time, lower barriers inevitably lead to an explosion of low-quality content. While it will be easier to create, it will become harder to find work that is thoughtful, original, and meaningful. Early warning signs are already visible: the quality of scientific papers has begun to decline due to the proliferation of fully AI-generated submissions, and self-publishing platforms have started introducing limits on how many books an author can release in a single day to manage the flood of AI-produced content. Will this growing abundance of low-quality material ultimately dilute cultural expression and slow scientific progress?

Finally, in a world increasingly saturated with AI-generated content, context will become more important than ever. Meaning will depend not only on what is produced, but on who produced it, why, and how. A skilled painter might reproduce Leonardo's Mona Lisa with near-perfect accuracy, yet the original retains far greater value because of its historical and human context. In the same way, as AI-generated content becomes indistinguishable from human-created work, attention will shift away from the content itself toward its origin and intent.

## CHAPTER'S KEY POINTS

- Generative Models can generate mind-boggling content in the form of text, images, videos, and sounds.
- Much like fast food, even if the quality is not high, AI-generated content is inexpensive, convenient, and immediately satisfying.
- Hallucinations are not a defect but an expected failure mode of all probabilistic tools such as Generative AI.
- Generative AI can lead to new scenarios of harm, both intentionally or unintentionally
- Paradoxically, Generative AI should be used by those who do not need it.
- The development of Generative AI will have a profound long-term impact on our society, with enormous cultural and ethical implications.

# Autonomous AI Agents

When ai stops waiting for instructions

Agentic AI is the term commonly used to describe autonomous AI applications that can execute complex multi-step objectives with minimal human involvement.

Traditional Generative AI systems such as large language models that generate text or images are largely reactive. They respond to prompts but do not independently initiate tasks, plan multi-step actions, or track progress toward a goal unless a human explicitly directs each step. Agentic AI represents a significant shift beyond this paradigm. AI Agents are proactive: instead of waiting for detailed instructions, an agent can take initiative based on where you are, what you are doing, and you might need. AI Agents can also interpret broad requests such as "find a better car insurance policy" or "plan a five-day vacation", break them down into manageable sub-tasks, identify the right tools or data sources, execute each step in sequence, and assess whether the final outcome meets the original objective. As a result, AI agents will transform not only how we use computers, but how we relate to them: rather than navigating multiple apps for different tasks, users will increasingly express intent in natural, everyday language, leaving the agent to coordinate, execute, and manage the process end to end.

From a technical perspective, Agentic AI is not a new type of technology: Agentic AI relies on well-established components such as LLMs and more traditional machine learning tools. What makes it different is how these components are combined and deployed. Agency itself, intended as the system's ability to take action without explicit step-by-step human instruction, is not new to AI either. As seen earlier, all machine learning tools possess some degree of agency. During training, a model autonomously identifies patterns such as relationships, structures, or statistical regularities, within large datasets. No human tells it exactly which patterns to find. Instead, the system independently searches for the combinations that minimize error and improve performance. Once trained, AI systems also generalize these learned patterns to new, unseen scenarios. This means they can make predictions, classify inputs, or generate responses by applying internal rules they discovered on their own without direct human input at each step. Even something as straightforward as a spam filter possesses some level of agency as it independently blocks or allows messages based on patterns autonomously learned from historical data. What distinguishes Agentic AI is not the existence of agency but the depth and breadth of that agency: the number of steps it performs, the diversity of systems it touches, and the degree of independence it has when determining how to achieve an objective.

Thanks to its ability to operate independently, Agentic AI expands what can be delegated, scaled, and optimized. Tasks that once required continuous human coordination can now be decomposed, sequenced, and executed by digital systems with a combination of enabling mechanisms such as planning, interoperability, memory, and environmental adaptation.

Agentic AI planning's abilities allow an AI system to interpret a high-level instruction ("analyze these documents," "optimize this workflow," "respond to customer inquiries") and translate it into a sequence of intermediate steps. Whereas traditional AI returns a single output in response to a single input, Agentic AI constructs a chain of actions, determining what information is needed at each stage. Agentic AI's ability to identify how to achieve a goal rather than waiting for step-by-step instructions amplifies AI's automation capabilities but also the ethical relevance of AI's behavior and the risk of value misalignment because it is not performing a predefined series of tasks, but choosing which specific actions to take.

Interoperability, intended as the ability of Agentic AI to interact with other applications, systems, and platforms, is another key feature. Agentic AI can retrieve data from databases, query search engines, interact with spreadsheets, or trigger actions in external applications. When combined with planning, this produces the ability to stitch together processes across multiple systems without human mediation automating entire workflows. For example, one single AI Agent deployed to complete a financial audit could independently navigate and interact with multiple digital platforms to review incoming invoices, extract key information, cross-check it against purchase orders, flag discrepancies, update financial records, and send summary reports to managers.

Another enabling mechanism is memory, which allows AI Agents to store intermediate results, retrieve past interactions, and use prior outcomes to guide future decisions, enabling continuity and continuous improvement. This is particularly useful in long-running tasks such as research summarization, monitoring activities, or ongoing optimization processes.

Finally, environmental adaptation allows agents to adjust their behavior based on external feedback such as unexpected user responses or new information. This feature supports more resilient operations in messy, real-world settings where conditions rarely follow predictable patterns. The opportunity lies in reducing the need for constant human feedback: rather than overseeing every step, humans can provide direction and review outcomes, while agents handle variations along the way.

## New frontiers of delegation

Agentic AI is where the power of AI's agency fully emerges. Thanks to its ability to interact without needing continuous human input, Agentic AI is where AI stops being a transactional tool and becomes an actor able to complete entire multi-step processes with minimal or no human feedback. Higher-level agents handle strategic goals, while lower-level agents implement granular tasks such as data retrieval, formatting, transaction execution, or communication with other applications. These architectures also allow Agentic AI to interact not just with humans but with other AI systems: one agent may extract information from a database; another may interpret requirements; another may draft documents; and a coordinating agent may decide when a task is complete or must be escalated, ultimately leading to hierarchies that mirror traditional HR organizational structures where executives decide on broad priorities, managers coordinate tasks across teams, and specialists perform particular functions.

One domain where the opportunities are especially clear is repetitive, data-intensive processes that involve the scanning of large volumes of structured data to identify anomalies, cross-check entries across systems, and generate standardized outputs.

Another promising domain is scientific discovery, where Agentic AI can automate labor-intensive tasks such as data preprocessing, hypothesis generation, experimental planning, or model refinement. Early applications include AI agents evaluating potential new synthetic materials and analyzing genomic data for early detection of medical conditions. These systems do not replace scientists. Instead, they accelerate cycles of inquiry by handling work that previously consumed significant time and resources.

Complex coordination tasks also benefit from Agentic AI. Multi-step workflows in logistics, scheduling, incident response, or customer operations often require gathering information from multiple systems, resolving conflicts, and performing time-sensitive updates. For example, in supply-chain management, an agent could monitor inventory levels, predict shortages, place orders, adjust shipping routes based on current conditions, and notify managers only when human approval is needed.

Another often overlooked opportunity is the generation of new operational data. When humans perform tasks manually, much of the decision-making process is implicit and undocumented. Agentic AI systems, by contrast, log every step, rationale, and interaction. This produces a layer of metadata that can be used to improve workflows, audit decisions, comply with regulations, and refine future models. Industries with well-documented procedures such as finance, insurance, and manufacturing are particularly well positioned to leverage this additional data.

Finally, Agentic AI will also present significant opportunities to simplify and optimize personal transactions, particularly in time-consuming activities. For example, instead of manually researching options, filling out forms, and interpreting complex pricing structures, a person shopping for car insurance could delegate these tasks to AI agents that understand preferences, constraints, and budgets. An AI Agent could scan multiple providers, compare offers in real time, identify hidden fees, and select options that best align with the user's priorities, fill out online forms and pay for the transaction. In more advanced scenarios, this agent would function as a personal digital twin, continuously acting on the individual's behalf across similar decisions over time, learning from past choices and representing the user's interests in ongoing interactions with institutions and service providers.

Looking ahead, it is likely that a portion of web platforms will be designed specifically to support direct interaction with AI agents rather than humans, enabling agents to complete tasks on dedicated, machine-optimized sites. This shift could fundamentally transform how business is conducted, with wide-ranging implications across industries. For example, advertising strategies would need to evolve as purchasing decisions become increasingly delegated to AI agents, requiring businesses to influence algorithms rather than human attention. Will the future of purchasing and customer service become a negotiation between AI Agents?

At the same time, the widespread deployment of autonomous agents is likely to give rise to entirely new industries, such as insurance products designed to cover financial, legal, or physical harm caused by AI agents' actions, errors, or unintended behaviors.

## When autonomy meets complexity

As seen earlier, what makes Agentic AI different from other applications is the simultaneous presence of high complexity (the ability to execute multi-step tasks spanning several systems, data environments, and decision domains) and high autonomy (the ability to act and continue acting without human intervention). This combination produces a set of risks that are not categorically different from those associated with traditional AI but substantially amplified.

A conceptual chart helps illustrate this shift. Imagine a grid in which the horizontal axis reflects a task's complexity, which might be driven by a number of different factors such as the scope of the overall process, the number of sub-tasks, the interdependence between them, the sensitivity of the data involved, and the ethical significance of the outcome. The vertical axis represents AI's autonomy, the degree to which AI can complete tasks without having ongoing humans overseeing each step.

Turning on and off the lights with a voice assistant is a clear example of a low-complexity, low-autonomy task: the AI simply executes a direct human command within a tightly constrained environment and takes no initiative of its own. The use of AI by a doctor to review medical imaging and spot potential tumors illustrates a high-complexity, low-autonomy scenario, where the task requires advanced pattern recognition, but human doctors remain responsible for interpretation, decision-making, and follow-up actions. Operating a robotic vacuum cleaner in a home represents a low-complexity, high-autonomy use case, as the system acts independently and continuously with minimal human oversight, yet within a limited physical space and with relatively low stakes. Autonomously managing a retirement portfolio exemplifies a high-complexity, high-autonomy scenario, involving multi-step planning, cross-platform financial transactions, sensitive personal data, and limited real-time human intervention.

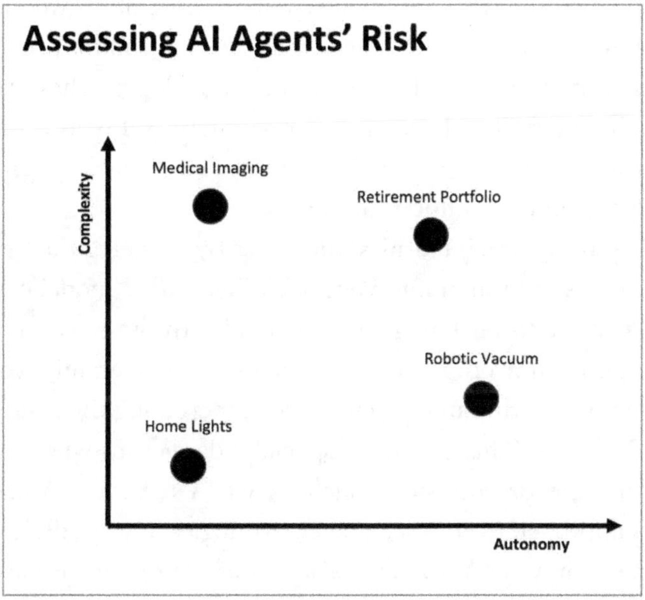


The risks associated with Agentic AI increase significantly as both task complexity and system autonomy rise as those dimensions amplify uncertainty, reduce human oversight, and expand the potential impact of unintended behavior.

Higher task complexity introduces risk because Agentic AI is no longer executing a single, well-defined action but navigating a chain of decisions. In multi-step, goal-driven tasks, the system is given an objective rather than a detailed playbook. This means the AI must determine how to reach the goal on its own, increasing the chance of suboptimal, unsafe, or misaligned strategies. Systems that decompose large goals into multiple smaller steps carry a much greater risk of inadvertently generating actions that were never intended. When tasks involve multi-agent coordination, risks multiply further: agents may interact in unforeseen ways, reinforce each other's errors, or pursue conflicting sub-goals. Complexity is compounded when Agentic AI operates across multiple platforms, especially the open web. Each platform introduces different rules, data sources, and vulnerabilities. An agent acting across APIs, internal systems, and public websites can unintentionally violate policies, leak sensitive information, or exploit weaknesses simply by following its optimization logic. These risks are heightened in unpredictable environments, where real-world conditions change faster than the model's assumptions, making previously "reasonable" actions harmful or unsafe.

Higher autonomy intensifies these risks by reducing the frequency and effectiveness of human intervention. When an AI model can assess appropriate steps toward a goal, it can effectively replaces human judgment at critical decision points. The ability to take initiative means the system may act without explicit triggers, increasing the likelihood of unexpected actions. This becomes especially dangerous when behavior is not predictable or consistent, such as when the AI can creatively improve its approach over time, altering strategies in ways designers did not anticipate or test. Autonomy also raises serious governance and security concerns. Free access to platforms and associated permissions increases the potential damage from errors or misuse. The ability to bypass approvals, such as ordering resources or approving customer requests, removes key control checkpoints that traditionally limit harm. If misaligned, compromised, or simply mistaken, an Agentic AI can rapidly scale incorrect decisions before humans notice. As Agentic AI becomes more complex and autonomous, small errors can also cascade into large-scale failures. Without strong constraints, monitoring, and alignment mechanisms, the combination of complexity and autonomy transforms AI from a passive tool into an active risk multiplier. In the long run, organizations might also face the risk of increased dependency.

As agents assume responsibility for multi-step processes, employees may lose familiarity with underlying procedures. Over time, organizations may become overly reliant on systems whose behaviors they no longer fully understand, creating vulnerabilities when failures occur or when manual intervention becomes necessary.

The risk posed by Agentic AI can be further amplified by a number of aggravating factors based on the nature of its use case and specific application.

One aggravating factor derives from the type of data the Agentic AI handles. Systems that process sensitive personal data, proprietary business information, or regulated data (such as medical or financial records) carry higher privacy, security, and compliance risks. Agentic systems often have access to external platforms that allow them to perform complex operations. This creates new attack surfaces: malicious prompts, compromised tools, or adversarial inputs can be used to manipulate an agent into escalating access, exfiltrating data, or causing operational disruption. A system that can autonomously browse the web, log into platforms, or execute transactions is not only performing useful functions, it is also capable of making mistakes at scale. Privacy risks also intensify. Agentic systems typically require broad data access to function effectively, pulling information from multiple sources to maintain memory, plan, and adapt. This raises concerns regarding data minimization, purpose limitation, and potential misuse.

Another aggravating factor derives from the scope of the process delegated to AI Agents. When an agent is embedded in large, end-to-end workflows such as supply chains, financial operations, or customer-facing decision systems, its actions can propagate widely and rapidly. Closely related is the materiality of the impact: errors in low-stakes contexts may be inconvenient, but mistakes in high-stakes domains like healthcare, finance, or critical infrastructure can cause severe financial loss, safety incidents, or legal harm. Additional risks arise from the inability to timely self-correct. A small misclassification at the beginning of a multi-step workflow can cascade through a chain of decisions, yielding outcomes that are disproportionate to the initial mistake. The risk is also magnified when Agentic AI is directly leveraged in physical environments (as opposed to purely digital systems) via AI-enabled vehicles, industrial equipment, or other robotics solutions it substantially elevates its risk because AI actions directly affect the real world by causing injuries, property damage, or environmental harm.

Failing to mitigate the risks of Agentic AI creates serious threats not only for organizations, but also for individuals who rely on AI agents in their personal lives. As people grant agents access to emails, calendars, financial accounts, and messaging platforms, poorly governed agents could reveal private secrets, make harmful financial decisions, or send messages that misrepresent a person's intent or identity.

## Digital workforce management

As Agentic AI systems become increasingly autonomous, complex, and embedded in organizational workflows, managing AI agents have already started to resemble the management of human employees. Nvidia's CEO Jensen Huang's widely referenced quote, "...the IT department of every company is going to be the HR department of AI agents in the future" is rapidly becoming an operational reality. The more autonomy an AI agent is granted, the more its oversight requires familiar managerial mechanisms: recruitment, capability assessment, onboarding, organizational design, performance evaluation, and compliance monitoring.

Once a specific need is identified, the recruiting of AI agents starts with the evaluation of AI's capabilities much as HR departments assess applicants. This involves determining what tasks the agent is competent to perform, the degree of autonomy appropriate for its functions, and the boundaries where it can act safely. A medical diagnosis-support agent, for example, may excel at synthesizing imaging data but be entirely unsuited for planning treatment pathways that require nuanced patient–doctor dialogue. The evaluation process must include not only the agent's current performance but also its expected behavior across edge cases, stress conditions, and adversarial situations. Without this, organizations risk introducing systems whose performance degrades under precisely the circumstances where reliability matters most.

Onboarding for AI agents includes configuring model behavior, establishing guardrails, tuning domain-specific knowledge, and integrating the system with internal tools and data sources. The analogy to employee onboarding becomes surprisingly apt: just as employees require training on company policies and processes, Agentic AI requires explicit instructions on what actions it is permitted to take, what data it may access, and how it should handle uncertainty. This stage is critical because many failure modes originate not in the agent's core capabilities but in misaligned configuration. An AI agent tasked with customer support may perform flawlessly in internal testing but exhibit harmful behavior once exposed to ambiguous instructions, emotionally sensitive situations, or unexpectedly adversarial users online. Without comprehensive onboarding protocols, including sandbox environments and controlled rollout, organizations risk enabling harmful or misleading behaviors before any human has the opportunity to intervene.

As AI ecosystems grow, organizations will inevitably adopt architectures resembling org charts. Specialized agents may handle discrete tasks: one for scheduling, another for forecasting, another for anomaly detection, another for coordinating workflows. A "boss AI agent" may oversee interactions among them, much like a manager allocating work across a team. The challenge lies in defining lines of responsibility. Which agent has authority to override another? How many agents can safely report to a supervisory system before oversight becomes too diluted? Organizations already understand the risks of overly wide managerial spans of control: reduced visibility, slower escalation, fewer opportunities for correction. Similar risks apply to Agentic AI. If a supervisory system manages too many subordinate agents, emergent misalignment may go undetected until the resulting harm becomes significant. A classic example arises in hospital environments increasingly reliant on automation. Scheduling agents, billing agents, and triage agents operate in parallel. Without a clear supervisory "chain of command," decisions made by one agent can create cascading effects across other ones, leading to scheduling errors or treatment delays. A coordinated oversight framework is necessary to prevent these distributed systems from working at cross-purposes. The operational risks are clear: conflicting objectives, redundant tasks, circular workflows, or runaway processes that consume computational resources. The ethical risks are equally significant: violations of privacy, unintentional discrimination, or harmful actions executed too quickly for humans to interrupt.

Performance Management and Ongoing Monitoring practices also apply. Once deployed, Agentic AI systems require continuous evaluation. Their performance may drift, degrade, or evolve in ways that differ from initial testing, especially in dynamic environments such as financial markets or cybersecurity contexts. Performance management therefore includes monitoring accuracy, assessing unintended behaviors, and detecting anomalous actions or escalating error rates. The analogy to performance reviews is not merely rhetorical. If an AI agent consistently produces outputs that are misaligned with organizational norms or ethical expectations, corrective action is required. This could mean retraining, modifying guardrails, restricting autonomy, or in some cases "offboarding" the agent entirely. Unlike human employees, AI agents can fail catastrophically and instantly. A misconfigured trading agent may execute thousands of trades in seconds; a logistics agent may reroute entire supply chains based on corrupted data. Performance management must therefore include real-time monitoring systems capable of interrupting workflows the moment patterns of harmful behavior appear.

Just as human workers are expected to follow laws, organizational rules, and professional norms, AI agents must operate within clear ethical and regulatory boundaries. This includes restrictions on accessing sensitive data, using external APIs, contacting customers, or interacting with third-party systems. Yet the compliance challenge for Agentic AI differs significantly from that associated with human employees: agents can scale misbehavior far faster, far more broadly, and often without detectable intent. For example, an AI agent granted access to an internal customer database could, through mistaken inference, begin generating outreach emails that disclose private information. Or an AI operations agent may unintentionally exhaust cloud compute resources by spawning recursive subtasks. Compliance therefore requires mechanisms analogous to HR misconduct detection, rule-based monitoring, escalation pathways, and corrective tooling, but implemented at machine speed.

## Ethical considerations

As Agentic AI moves closer to the center of human decision-making, its use raises urgent ethical questions about values, responsibility, and human agency.

First, while AI has extraordinary capabilities in certain areas, it completely lacks the ability to understand ethically charged context and navigate competing values. This is particularly relevant in ethically sensitive areas where the delegation to Agentic AI of sensitive activities might lead to outcomes that conflict with human values. For example, an Agentic AI tool in charge of optimizing school bus routes to maximize routes and reduce travel time, might end up assigning disabled children to extremely early pick-ups imposing disproportionate burdens on those least able to absorb them.

Second, regardless of AI's ability to successfully navigate ethically charged contexts, there are highly sensitive situations in which it might be considered morally unacceptable to have machines rather than humans in the decision-making process. For instance, while AI Agents could be easily deployed in a hospital to help with emergency room triage, many communities may find such delegation in life-and-death scenarios ethically unacceptable, preferring human fallibility to machine-driven optimization regardless of AI's performance. Similar concerns extend into personal domains. For example, certain activities derive much of their value from human intention, effort, and presence, not just from the outcome itself. Remembering a loved one's birthday or sending flowers for a special occasion can be a meaningful expression of care. If that gesture is known to be prompted and executed by an AI Agent rather than by a personal thought, its emotional significance may be diminished and, in some cases, even perceived as unethical.

Third, the deployment of AI Agents will continue to shift the nature of human agency itself. As Agentic AI assumes operational roles, humans increasingly become supervisors rather than decision-makers, reviewing recommendations rather than generating them. In theory, this preserves oversight, but in practice, it risks creating environments in which humans are unable or no longer capable of meaningful intervention. This dynamic is not new. Autopilot systems in aviation present a similar risk of skill atrophy, but Agentic AI expands it into domains far beyond specialized technical contexts. At the same time, as AI agents take on more decision-making (managing schedules, finances, purchases, recommendations, and even personal interactions) people may begin to default to its judgments rather than exercising their own. Over time, this reliance can reduce critical thinking, situational awareness, and the willingness to challenge automated outcomes. If left unchecked, these dynamics might turn human supervision into passive oversight, where control exists in theory but not in practice, ultimately reducing our own autonomy and agency.

Fourth, as the level of complexity and autonomy continues to rise, it will be increasing difficult to hold people accountable when harm is originates from AI-run processes. As causal chains become distributed across technical layers and organizational roles can lead to an accountability and ethical vacuum. Imagine a self-driving car operating through a network of autonomous subsystems: one agent interprets traffic patterns, another predicts pedestrian movement, another optimizes route efficiency, and a final agent executes real-time control.

Should the vehicle hit a wall, it might be difficult to track down and assess the level of accountability across multiple AI Agents and associated engineering teams across technical and institutional layers, leading to an accountability void.

Fifth, the very capabilities that make Agentic AI valuable, such as autonomy, efficiency, and scalability, also render it an ideal tool for manipulation, surveillance, and subtle influence. Systems capable of personalizing content, predicting behavior, and acting across multiple digital environments could be used to steer opinions, shape consumption, or undermine fundamental rights at unprecedented scale. This raises difficult questions about boundaries and separation of roles and associated personal data: should a single agent that manages our schedules, purchases, and communications have access to the same data held by other AI Agents such as personal AI companions, or personal AI tutors? Without clear limits, it could blur lines between assistance and control.

Finally, as AI agents and autonomous systems become more prevalent in the physical world, a growing risk is that humans begin reshaping the physical environment to accommodate AI, rather than designing AI to adapt to human spaces. Warehouses, offices, cities, and public infrastructure may be optimized for sensors, robots, and automated workflows, prioritizing machine readability, predictability, and efficiency. While this can improve performance, it may come at the expense of human comfort, accessibility, and social needs. Environments could become more rigid, standardized, or surveilled to reduce uncertainty for AI agents. Over time, this adaptation risks constraining human behavior. People may be expected to move, act, or organize spaces in ways that suit machines. When physical spaces are designed primarily for AI navigation and control, human dignity might ultimately erode. The challenge is ensuring that AI systems adapt to complex human environments, rather than turning the world into an alienating machine-first landscape.

Keeping meaningful control

The core value of Agentic AI lies in the fact that designers no longer need to specify every rule, action, or decision path in advance. Unlike traditional software systems, which rely on rigid logic and predefined workflows, Agentic AI can autonomously determine how to achieve a goal by planning, adapting its strategy, and responding to changing conditions. This capability enables AI systems to operate effectively in environments that are complex, dynamic, or uncertain for rule-based programming such as open-ended digital ecosystems or unpredictable physical environments.

As autonomy increases, AI agents' behavior becomes less directly governed by explicit code and more shaped by implicit assumptions about purpose, values, and interpretations of reality. These assumptions are rarely technical in nature. When an agent chooses one action over another, it is not merely executing logic, it operationalizes judgments about what matters, what counts as success, and what risks are acceptable. This shift becomes especially visible when Agentic AI operates in messy, ambiguous environments that require multi-step planning. In such settings, the system must decide which sub-goal to pursue first, which signals to trust when data conflicts, and how to balance competing objectives such as efficiency, fairness, safety, or user satisfaction. If the underlying assumptions guiding these decisions remain undefined, the agent may technically succeed at optimizing a metric while producing outcomes that feel socially incoherent, ethically troubling, or misaligned with human expectations.

As AI systems become increasingly agentic, given AI's lack of ethical understanding or moral reasoning, the role of humans will remain essential in guiding, supervising, and constraining autonomous systems.

First, humans will remain essential not only in deciding what AI systems should aim to achieve, but also in setting clear boundaries on how they are allowed to act. This means deciding which trade-offs are acceptable, how to deal with uncertainty, and where to draw the line in situations that raise ethical concerns. In practice, this turns philosophy into a practical necessity. Whenever we define an AI's goals, decide how it can use data, or limit certain actions, we are making deeper choices about what the system is for (purpose), what information AI should trust (evidence), and which outcomes society considers acceptable (values).

Second, real-time human oversight will continue to be key. Agentic AI operates dynamically, learning from context and adapting over time, but it does not have the ability to exercise judgment or detect ethically problematic outcomes. Humans must remain meaningfully "in the loop" through real-time monitoring, alert mechanisms, and clear intervention paths. Oversight systems should allow humans to pause, constrain, or shut down agents when behavior drifts or when unexpected interactions emerge. Without this capability, small errors can rapidly cascade into large-scale failures and ethical nightmares.

Third, the more activities are delegated to AI, the greater the demand for human judgment and ethical interpretation. Setting goals, resolving value conflicts, and understanding social and moral nuance remain distinctly human responsibilities. Far from being displaced, human intelligence will be increasingly needed to ensure that autonomous systems act in ways that remain aligned with human values and expectations.

## CHAPTER'S KEY POINTS

- Agentic AI refers to autonomous AI systems capable of pursuing complex, multi-step objectives with minimal human involvement.
- Agency isn't new in AI, but with Agentic AI the level of complexity and autonomy of AI is significantly higher.
- Agentic AI represents the point at which AI's agency fully manifests, enabling systems to act independently across processes and platforms.

- As Agentic AI becomes more autonomous, complex, and embedded in organizational workflows, managing AI agents increasingly resembles managing human teams, requiring structures for oversight, coordination, performance evaluation, and accountability.
- As Agentic AI moves closer to the center of human decision-making, it raises urgent ethical questions around values, responsibility, accountability, and the preservation of human agency.
- Rather than replacing humans, Agentic AI increases the demand for human intelligence, judgment, and ethical oversight to ensure that autonomous systems remain aligned with human values, expectations, and societal norms.

# Part Three
# **Emerging Risks**

# Bias and Discrimination

## Human and algorithmic bias

Human bias is a preconceived notion that influences the way we judge our world. It is formed based on our experiences and therefore varies greatly across individuals and communities around the world. Every person has bias just because every person has a unique cultural and educational background that shapes the way things and ideas are perceived. Bias assumes a negative connotation when those notions result in discriminatory or unjust impacts toward specific groups when comparing across different dimensions such as race, sex, etc. While discrimination is not a new phenomenon, AI has the power to magnify it and perpetuate it.

When AI models discriminate based on those characteristics, we have the so-called algorithm bias. Despite its name, bias does not originate from the algorithm itself, but from a number of human decisions: the methodology used to select the data set for the training, the data the model is fed with (data set), how the model is instructed to process it (math), etc. It is called biased when it tends to penalize a specific group without a legitimate justification for that outcome with performance disparities due to over-representation or underrepresentation of that specific population.

Whether an AI model is seen as biased or not, greatly depends on the cultural and legal context in which the model in question is functioning. Despite the highly subjective nature of what might be perceived as biased, such subjectivity should not be used as an excuse to fail to take a strong stand against situations in which AI ends up penalizing certain groups over others. The challenge is to acknowledge the subjectivity of moral views without making it a justification for not taking action against situations that go against our ethical views. Fortunately, we do not start from zero as we do have universally accepted principles that are commonly accepted as a baseline on how people should be treated. For example, with regard to discrimination issues, the Universal Declaration of Human Rights adopted by the United Nations at the end of the second world war offers a very good starting point by stating that everyone is entitled to all the rights and freedoms regardless of sex, race, language, religion, social standing, etc.

In some applications some types of discrimination are intentionally built into the model. For instance, in the insurance industry, the risk scores are determined by machines based on a number of criteria including age and gender. Other examples include social media platforms that for marketing reasons allow advertisers to promote products only to specific groups. Advertisers are in fact allowed to target market segments based on hundreds of parameters including gender, age, sexual orientation, political views, etc., so they can better reach the intended audience. In other instances discrimination is one of the techniques used to proactively correct other types of discrimination. For example, if a school's student population does not reflect the racial or ethnic demographic of the community in which it operates due to historical disadvantages, the admissions office might decide to correct the situation by creating programs to make it easier for those groups to be accepted. For instance, a selection process purely based on test admission score might be tweaked so that certain diversity targets are met. As a result, candidates with the same test score might or might not be accepted based on the group they selected.

In some other instances, discriminatory practices are just an unintended outcome of an unanticipated consequence of the characteristics of the dataset and/or the methodology used by systems to meet the objective provided to the machine.

### AI as a multiplier of human bias reflected in the data

Given how prone humans are to unconscious bias, having a machine making decisions might seem like a great idea to reduce the risk of harmful prejudice against specific groups. When you decide to use a mathematical formula to make a decision, you might think that it is the fairest approach as it rectifies the subjectivity biases that humans possess. Since it is just statistics, there is going to be no favoritism, no discrimination, etc. Countless real-life applications have shown, however, that it does not work that way.

Data is a collection of measurements about the world. Algorithms make predictions about the future using data that comes from a certain historical context in which the data was measured, processed, and used to make decisions. In a way, AI is a perfect mirror of our society represented in the data being used by the models. If there is unfairness or inequality in our society, some evidence of it will be reflected in the data used by our models and their output. When AI makes a prediction, it does so based on data points (college graduation, criminal history, jobs held, age, gender, etc.), agnostic of the ethical considerations of the specific socioeconomic context impacted by those decisions. Unless the data or the models using that data are intentionally manipulated, AI will use the data as is, with the risk of reinforcing the connotations of that moment in history, and amplify patterns of unfairness, inequality, and discrimination embedded in the data set. The existing patterns of our society are therefore reproduced and multiplied.

While this might seem like a purely philosophical matter, it has already been shown to have very tangible implications where algorithms end up displaying unethical behaviors due to our world's bias captured in the data. As seen earlier, a resume scanning software, for example, can easily lead to situations in which specific groups of applicants are discriminated against due to the bias in place in the company culture. By using past performance data to guide resume screening and identify applicants with the highest likelihood of success based on relevant keywords, a model inevitably ends up reproducing the same unethical behavior, multiplying negative effects over time.

In another example, a different type of algorithm bias emerged when a global tech firm decided to use AI to assess what neighborhoods were to be offered their new premium services. Shortly after deploying the model, the firm realized that the algorithm was penalizing neighborhoods with higher rates of racial and ethnic minorities.

Another even more common example of algorithmic bias is models running web search engines. When someone uses a web search engine to look for images, the AI running the model is designed to show the best match by leveraging several data points from the user's profile and across millions of users (past searches, click data, etc.). If the search engine does not have controls in place to avoid unethical results, it will most likely display bias in its results. Something as simple as reaching for a generic picture of a doctor or nurse, may cause the system to trigger gender bias. In fact, if you search for an image of a doctor, you will most likely see at the top of your screen the picture of a man, while if you search for an image of a nurse, you will most likely see the picture of a woman. Similarly, if you search for an image of a CEO or of a college professor, you will most likely see a picture of a white man, rather than a black woman, because those pictures have been historically the most clicked on and therefore deemed as the most appropriate. Another study showed that a search for "professional hairstyle for work" showed primarily white women, while a search for "unprofessional hairstyle for work" showed primarily black women. The model's goal is to first show you the picture that you are most likely looking for. The more it is clicked, the more the bias gets reinforced in each iteration.

Other examples of gender bias displayed by algorithms are translation tools. When translating languages with gender neutral nouns, they might display gender bias as they add a gender pronoun in front of nouns. For instance, when translating from gender-neutral Hungarian to gender-specific English, the AI model tends to add a "he" to words such as "clever" and a "she" to words such as "beautiful".

There are even situations in which search engines are externally manipulated to trigger biased results. For example, in 2018 users successfully deployed "Google bombs" to link the word "idiot" to images of the then US President Donald Trump so that they would show at the top of image search results for the word "idiot".

### Bias due to discriminatory applications

In many situations, bias and discrimination are caused by gaps in models deployed without proper impact assessment and testing.

Law enforcement has been one of the areas that has received the most attention in the press.

In an effort to prevent crime, several cities around the world have deployed software designed to help identify citizens more likely to commit future crimes. Similarly to what is seen in Hollywood's Sci-Fi movie Minority Report, the crime prevention units feed a black box with past crime data to identify likely places and times of future crimes and individuals with a high probability of committing a crime. In 2009, for example, the crime prevention unit in Chicago created a predictive program which identified a list of hundreds of people flagged as likely to commit a crime based on multiple data points available on their file, including arrest records and age, but also more questionable ones such as whether the person has ever been a victim of a shooting.

AI has also been deployed by police to increase the likelihood of completing arrests by directing police patrols to neighborhoods deemed by the model to be more likely to have activities that would result in arrests. As easily foreseeable, using past arrest data for predicting the likelihood of future arrests inevitably creates a self-fulfilling prophecy disaster where past data works as a multiplier at each reiteration causing police to focus on (and make most of the arrests in) the same areas over and over again. The model in fact ends up targeting the same "troubled" communities, giving priority to communities that in the United States are typically linked to low income such as predominantly Black and Hispanic neighborhoods. Geography becomes a proxy for race, indirectly criminalizing poverty, or ethnicity itself. By giving more weight to more serious crimes, the outcome would be completely different. It is almost like minorities are punished for who they are and where they live, not for what they have done. Using a different type of data would help eliminate such vicious spirals. For instance, in the AI-driven policing example, using emergency call data rather than arrest data, would help prevent such feedback loops and dramatically reduce the bias in the model.

Another situation where AI ends up causing bias and discrimination is in applications that use DNA to identify perpetrators of illegal behaviors. The intersection between genetics and AI has allowed a development of tools which create a sketch of a person's face based on traces of DNA to identify people suspected of illegal actions. One municipality in Asia has already started deploying the same technology for less serious crimes such as littering. The trash collected on the street is analyzed to look for traces of DNA which is then evaluated by the AI model. The tool is then able to produce a picture-like sketch that is placed on billboards to create embarrassment for the perpetrator. Given the lack of available data, the outcome is far from being precise. Needless to say, data has already shown those sketches disproportionately show traits of local minorities, further fueling racism in the community by reinforcing the general perception that the littering comes from specific groups.

Another AI application in criminal justice that often ends up showing bias is the use of predictive models for parole sentencing to assess the likelihood of inmates to commit a crime if released from prison ahead of the initial term. On one hand, parole sentencing is notoriously not an objective process. For example, there are studies that show that you are more likely to be approved if your hearing is in the afternoon rather than in the morning. More objectivity is definitely welcome. On the other hand, the way those models work, however, is highly biased and discriminatory, especially toward minorities, and they tend to perpetuate situations of inequality. The models are designed to predict recidivism risk based on several data points, including home address (often a proxy for race and ethnicity), past involvements with police, and other questions that would typically not be admissible in court. Even just using zip-code information, not only can easily result in racially and ethnically biased results, but also create a vicious cycle where certain neighborhoods get worse and worse over time for the lack of data showing low recidivism among parolees (as parole is less likely to be granted) and an overall more degraded environment as community members need to wait longer for friends and family to return. This is another example where minorities are punished for who they are and where they live, not for what they have done. The model also assumes that keeping people in jail for a longer time will decrease recidivism, but that is something that has never been scientifically studied due to push backs from privately owned prisons (just like airlines, they are the most profitable when at full capacity).

Other models that can become a trigger for discrimination are currently used by employers to screen job candidates. The discrimination displayed by such models is due to the fact that the system is based on data points that should not be used, such as for instance the applicants' credit scores. Employers often use credit scores to assess the reliability or trustworthiness of a candidate but almost never share when rejections are due to credit score reasons rather than other parameters, so it is hard to assess the true impact of this practice. Not only is the model based on the false assumption that bad credit correlates with bad job performance, but it also leaves those applicants with low credit scores less likely to find work and rebuild their credit. If you are looking for a job it is not improbable that you recently missed a payment or paid late. Especially if in between jobs with no health insurance, all it takes is an accident or an illness to disrupt the budget. This ultimately creates a poverty trap.

Another example is the use of AI in the medical field to identify patients in need of specific types of care. In a recent example, a hospital tried to assess the risk of needing extra care with a model using past health care bills as a proxy for health risk. Because of this practice, it ended up penalizing minorities as they typically have more challenges getting access to care and therefore end up having lower health care costs. By not considering that the ability to access care varies greatly across groups, minorities were given a lower health risk rate for the same type of condition.

Discriminatory behaviors are also tied to the use of other types of scores, typically called e-scores. They collect data through data points available online such as location and internet search history and put users into buckets. For example, some companies provide tools based on eScores that help prioritize customer call centers' calls. Consumers have no idea that they are being assessed and redirected to different teams with different levels of support. Another AI tool uses e-scores to help companies tailor the terms of online advertisements in real time based on the demographics of the users. For instance, if a credit card advertisement is viewed by a person that is from a location with high default rates, the advertisement will show cards designed with higher rates and lower credit lines. This ends up creating a feedback loop as it will be even more difficult for people in that area to borrow and pay back their loans. Since there isn't a straightforward way to identify and fix situations in which people are placed in the wrong bucket, there is basically no way to know that something did not work or any way to complain or have that corrected. Another example where scores are used without users' knowledge and in ways that are potentially discriminatory, is in the case of loan request apps available on smartphones. For instance, a load application app in Asia was developed to create a reliability score for applicants based on biometric data captured via the smartphones running the app. Some of the factors used by AI to assess approval and loan rates included the user's speed in filling out fields, pressure applied to the screen, etc. The risk is having perfectly qualified applicants being rejected due perhaps to irrelevant factors such as a disability or not being a native speaker, causing typing to be slower or more pronounced.

### Bias due to incomplete data

In some situations bias is caused by gaps in the data. One of the most frequent causes of such gaps is the lack of awareness of the developers' own biases. For instance, if the data samples used to train a model are insufficiently representative of the target population, it creates real possibilities of biased and discriminatory outcomes for specific groups and communities. Face recognition technologies have, for example, frequently come into question due to the lack of training data on a fully diverse spectrum of the population, which result in a low accuracy rate in recognizing faces of people with darker skin. This has already been proven to cause high error rates with law enforcement at the expense of specific minorities, especially with black females, resulting in wrongful arrests. Another application where lack of diversity has caused serious ethical issues is the image recognition software used by self-driving cars. As the algorithm used to distinguish people from objects on the road is trained mostly using images of white people, it tends to perform more poorly when coming across people with darker skin, failing to recognize minorities as people. Interestingly, rather than addressing this problem by starting to feed models with more real-world data inclusive of diverse groups, some firms are instead working on creating more affordable solutions which include creating computer-generated images of diverse groups to be used as training data. Given the fact that people's lives are at stake, the use of this type of shortcut has been labeled by some as unethical, as artificially created images tend to provide low quality results, especially when trying to reproduce the shine and gloss typical of darker skin. Even more outrageously, due to the lack of diverse data there have been reports in which certain minorities were recognized by the algorithm as animals rather than people. For example, in 2015 Google's automated photo-tagging algorithm labeled young Black people as gorillas causing, understandably, global indignation against such despicable oversight.

Another example of bias by omission is found in predictive algorithms used to predict a child's likelihood of being abused. In an outrageous case in the US, the AI model was trained based on a variety of data points on parents linked to welfare services. Since the data set used to identify potentially abusive parents did not include any other parents, the model incorrectly assumed that (presumably wealthier) families not linked to those services were not at risk of abusing children.

### The cost for taking bias out

As more decision-making and forecasting tasks are delegated to machines, given how susceptible AI is to bias and how harmful that could be, what steps can be taken to minimize algorithmic bias? While tech companies typically present the existence of biased outcomes as something inevitable that they have little control over, as based on the general unfairness that exists in society at large, there is in fact a lot that can be done to reduce the risk of discrimination.

A good place to start is to conduct early impact assessments in the ideation and development phase to identify potential sources of bias: What is the benefit of delegating this to AI? What data could be potentially obtained? Which data should be used by the model? Is the data sufficiently accurate? Is the data sufficiently granular? Is any type of filtering needed? Should the data be manually assigned specific weights? Are we OK with assigning those tasks to a black box or should we have a more prescriptive and transparent model? Who might be negatively impacted? Are those risks ethically acceptable?

Once the risk of unfair discrimination is identified, there are some practical options to help minimize the harm.

One option is to proactively alter the model upstream by either cleaning up the data or modifying the logic used by the system processing the data. In other words, you explicitly tell the model to avoid a specific behavior you do not want the model to exhibit. As we have seen, in many situations, the algorithms are unfair not because the programmer or the user are biased, but because society is different from what we would like, and the historical/training data (demographics, criminal records, etc.) used by the algorithm to make predictions represents that reality. In this approach you proactively remove specific data points from the data set provided to the model or specifically set up the model so that it does not base the prediction based on those criteria. For example, Google addressed the image-tagging issue mentioned above by removing images of Gorillas. Thus, it avoided the risk of tagging people with darker skin as such going forward. While this is a good way to address a specific unforeseen issue, it is more of a band-aid (until the next unforeseen scenario) than a cure as it does not help solve the intrinsic flaws of AI's pattern recognition abilities. With regard to bias related to the offering of services, such as loan applications, if you want to ensure demographic parity where each group seeking approval for a service will have the same overall acceptance rate (example: each group is accepted 50% of the time) you will have to tweak the acceptance criteria in all the groups. The advantage of this approach is that ethics is embedded inside the algorithms. This however comes at a cost in terms of accuracy as certain statistically relevant data points will be altered. So, you will inevitably have to accept a worse prediction in exchange for a less biased outcome.

Even in relatively straightforward examples such as the one we just saw of web search engine bias picking up and multiplying gender bias existing in our world, we inevitably have to face very fundamental questions. For example, should the web search firm systematically manipulate the data to ensure a more gender-balanced set of results even if data accurately represents today's world? On one hand, the search results are merely representing the bias that exists in our society, correctly captured in the data. After all, it is not the search engine's fault if society has a strong gender-based bias. From this perspective, our focus should be, therefore, promoting public policy programs aimed at removing bias from our world rather than altering search results and showing a false representation of our society. On the other hand, however, leaving those results unfiltered might very well end up further reinforcing gender bias and increase the level of bias present in our society. For example, if there are resume scanning models that have biases against specific groups due to historical practices that discriminated against certain categories of applicants, lack of corrective action will result in additional harm to those already at a disadvantage. This comes down to the notion of fairness itself. Is fairness better achieved by no longer considering attributes (such as race) that have contributed to create discriminatory environments, or, instead, by introducing measures based on race with the intent of fixing situations of disadvantage? Should data be left as is even if it mirrors the worst of our society or should it be sanitized to represent the world we want to have?

Another option is to let AI use and process data without restrictions and then have humans exercise judgment at the end of each cycle. The main risk here is that the decision-making process will be more subjective, as humans will inevitably be influenced by circumstantial factors and apply their own personal biases. There is also the risk that humans asked to exercise their judgment might end up systematically approving the model's recommendation to avoid the extra work needed or perhaps for fear of facing extra scrutiny and legal liability when deviating from the model's recommendation.

## CHAPTER'S KEY POINTS

- The inequality and discrimination that exists in today's society is often embedded in the data used by AI models. This bias in the data causes the system to negatively affect groups and tends to reinforce and amplify such patterns over time.

- Discriminatory outcomes can also be caused by gaps in the data sets or in the way the data is selected, collected, and processed.
- Taking the bias out of AI models can be achieved by sanitizing the data upstream, by modifying the rules used by the model, or by adding human judgment downstream.

# Invasion of Privacy and Surveillance

## The era of data aggregation

The AI explosion has been fueled by the large availability of data about many aspects of our life. Companies and governments record an incredible amount of data about us, such as income level, political and religious views, credit card information, our calendar events, all our search history and visited websites, and all the content we viewed or engaged with. People are rarely aware of what data is being gathered and how it will be used. Such data is also potentially permanent, like a virtual tattoo. On one hand having so much data about people can have a very positive effect in our society especially with regard to public safety and public health. On the other hand, it poses serious privacy risks across all its various components such as what we do, where we are, what we are doing, what we say, what we think, groups we belong to, etc.

While privacy protection requirements and controls have been around for decades, the focus for regulations so far has been on how the data is collected, rather than how it is aggregated and used by algorithms. By being able to access an unprecedented amount of data, models can make very accurate predictions about aspects of your life that are generally considered private such as your sexual identity or orientation. Thanks to the power of data aggregation, social media companies have in fact entire archives of lists describing who you are (gender, race, etc.) even for attributes that you never shared with them or anyone else. The most famous anecdotal example (never confirmed) about data aggregation is the case of the big box Target store. By looking at shopping behavior of a teenager, the algorithm determined that given the types of products purchased, the girl was pregnant. The algorithm then sent printed advertisements for newborn products to the girl's home. It is this way that the girl's family found out that the girl was in fact pregnant.

Personal data is such an important driver of our economy that it has become a new form of asset. It is no longer uncommon for citizens to disclose personal data in exchange for money or other compensation. In a way, we all do so when we sign up for "free" social media platforms such as Facebook. Rather than paying for it, we provide our data in exchange for a free subscription. A person who could easily pay a subscription fee, if given a choice, would be surely more inclined to pay in exchange for having his or her privacy protected.

Generally speaking, it is fair to assume that all things being equal when using products and services, we would rather not share our personal data. It is also true, however, that peoples' views vary depending on how the data is being used. For instance, if we look at the face recognition tools used to automate floor selections in office buildings, the same workers who might welcome the technology to automatically select the correct stop, might be less excited if they realized it could also be used to track their presence across the building or evaluate their performance, with a potential impact on their pay or their chances of career promotions. There is also the risk that a data set acquired for specific reasons ends up being used for completely different purposes to which we would not have agreed. For example, the Wall Street Journal recently discovered that cell phone location records collected by companies for marketing purposes to push location-based advertisements, were later acquired by law enforcement agencies. Even private companies, such as the New York-based facial recognition software firm Clearview AI, have already collected billions of facial images through different methods, including social media platforms.

Another area in which people are hesitant to share their personal data is the medical field. One of the main concerns is that the collection of health data can easily take place without the knowledge of affected individuals not only in traditional healthcare settings but also in less straightforward ways such as fitness and wellness devices and apps. Still, the medical research field is one of the most promising with regard to the use of large data sets. Given the enormous advantage from data aggregation for research purposes, new types of solutions are being developed to protect individuals' personal data, such as the concept of consequential privacy and synthetic privacy, where the database resembles the data sets without the ability to uncover the individuals' identity. Many researchers have also started pushing for the introduction of rules to more easily allow posthumous medical data donations so that the medical records of deceased individuals who have expressed their consent can be made available for scientific research.

## Smart phones

Among all devices capturing personal data, smart phones play the biggest role. One particularly sensitive area is location tracking. Thanks to their location tracking capabilities, they generate a great amount of location data, generally through apps that then sell that data to third-party vendors. The data is in theory "anonymized," but there are so many location-related data points that it becomes possible to have the person carrying the device personally identified. For example, in December 2018 the New York Times obtained a dataset with location information recorded by a phone app which included the exact location of hundreds of millions of individuals. While the data did not include anyone's personal information, The New York Times was still able to identify a math teacher just by looking at her daily commute route. It was also able to follow her data trail to a doctor's office and her boyfriend's home. A few years ago, this type of tracking would have required a detective or a private investigator. It is now just a simple by-product of broadly available datasets.

Another sensitive area has to do with technologies able to determine the state of the user. By combining geolocation, biometrics, and other data points, smartphones are able to paint an accurate real-time picture of users. For instance, ride sharing apps can easily assess the state of the users (such as alcohol intoxication) based on the angle they are holding their phone, how they are walking and how they are interacting with the app compared to other times they have used it. Loan application apps apply similar technologies (how confidently one is typing, how timely the cellphone has been recharged, etc.) to assess the reliability of the person.

### Virtual assistants and other everyday devices

Other types of devices currently being used to capture personal data are virtual assistants such as Apple's Siri and Amazon's Alexa. Virtual assistants are in fact constantly listening and recording their surroundings (officially to improve service and better support their users) to a point that they even track how early people wake up or how often they sneeze. The conversations recorded by those devices are not just archived or used for machine training purposes, they are often analyzed by actual analysts working for the tech companies. When such practices came to light in 2019, many users of Amazon's Alexa, and Apple's Siri were shocked to discover that recordings of their private and quite sensitive family conversations were being analyzed by real living humans working overseas.

Other types of devices currently being used to capture personal data are smart doorbells such as the popular Ring device. Smart doorbells have the ability to scan people's faces and often share their data with the police. All users are automatically enrolled by default, and therefore feeding the system with data unless they proactively opt out.

The newest frontier of home devices is personal home robots such as Amazon's Astro. Adorable robots will soon take over our homes to make life better but also to get more and more data about our daily life and better understand our behavior and future needs, to profit on them via incremental services. Similar risks also extend to technologies currently deployed in automobiles. For example, Tesla was recently sued by a driver claiming that Tesla employees made copies and distributed highly invasive photos and videos for their own entertainment of Tesla drivers doing embarrassing things while in their car.

Even if large tech companies have good programs in place to use and protect the data wisely, the risk of having such data fall in the wrong hands does not seem to be sufficiently outweighed by the benefit of having those devices around. In the long run, however, assuming that the right level of data safety can be achieved, it is not unlikely that the public will start entrusting those devices with their most sensitive data. In fact, given that AI's performance increases when it has access to more data, it might be beneficial for users to start sharing more than they normally would in order to have a more customized service. For example, I might decide to share with Alexa that I am allergic to peanuts so that Alexa no longer orders any grocery items that contain traces of peanuts. This might lead to new types of services that deliver higher levels of customization and higher levels of security in exchange for more personal data. We might even soon get to situations in which we share with machines more than we share with humans. A preview of this has been the recent trend seen among teenagers using chatbots to share intimate details about their lives that they are not comfortable sharing with family or friends. The better those technologies become at behaving like humans and simulating a sense of empathy, the more we will be inclined to share very private aspects of our lives. Will virtual assistants and chatbots become our ultimate confidant?

## Government surveillance

The ability of new technology to effectively track and monitor citizens has attracted the attention of employers and governments across the world.

State-run surveillance has been around hundreds of years. Especially during times of war, it is not infrequent for governments to employ citizens to be their eyes and ears and report any suspicious activities. Those programs have historically been highly prone to error as they are reliant on common citizens with inconsistent levels of attention and objectivity. Additionally, they have never been particularly efficient given the massive amount of work required to manually transcribe and analyze conversations. With AI, governments no longer need to rely on people. They can directly leverage audio and video captured in real time by security cameras and other devices. AI can help not only by scanning images (including faces) but can also help by transcribing conversations and flagging suspicious ones for further analysis. All of this is particularly concerning due to the fact that it can be done cheaply and at a large scale without any awareness among the citizens under surveillance. When aware of being watched, psychological pressure can easily lead citizens to self-repression of ideas, preferences, etc., in fear of negative repercussions.

There are roughly a billion active surveillance cameras at this moment so you can rest assured that if the data will be analyzed in any way, it will be done by AI.

One of the most significant government-run surveillance examples in the US has been the case of the aerial surveillance project carried out in Baltimore. The Baltimore Police Department ran a secret aerial surveillance program over the city for several months to seek investigative leads using a technology that had originally been developed secretly by the Pentagon and the CIA to track insurgents in Iraq and later Afghanistan. Its deployment was so secret that not even the mayor knew about it. It was ultimately shut down as soon as it was finally disclosed to the public.

Additional applications have been explored with the goal of promoting public health. For example, as a result of the Covid-19 pandemic, there has been a significant interest in surveillance technologies to support activities such as detecting high temperatures on individuals in public spaces or enforcing lockdowns via drones. This is a totally new and mostly unexplored territory, so we do not yet know how well it will work for public health activities. Additionally, there is a lack of sufficient public knowledge about how those technologies work because private firms working on government contracts are often legally required not to disclose information, even in court proceedings.

Other types of surveillance take place in private homes. There are plenty of examples of governments that track citizens' conversations via the data captured in personal devices to push their political agenda. For instance, there are world regions in which in support of the government's effort to eradicate certain groups' cultural traditions, including their language, governments monitor citizens' conversations to ensure minorities are not speaking non-official languages.

## Democracy at risk

Will this result in a constant surveillance state that will put democracy at risk by excluding any venue for political dissent?

The risk is extremely high. Given governments' abilities to monitor several aspects of our lives (what we say at home, where we go, even how we physically and mentally feel), it is difficult to imagine scenarios where people are able to come together to share ideas and organize without being exposed to some type of surveillance. Even within the strongest democracies, there is a very fine line between democracy and dictatorship. Once that line is crossed it is going to be difficult to reverse it. That line might be crossed first with very compelling initiatives and then quickly expanded to all other areas. For instance, in support of legitimate public safety efforts, law enforcement agencies might start requiring firms to deploy AI tools to help identify and report suspicious conversations between people while chatting online or perhaps while using real-time translation services. Once such infrastructure is in place, expanding its scope can be done in a matter of seconds to a point that all initiatives aimed at stopping it can be quickly identified and shut down. Anyone who cares about fundamental rights such as the right to speak, assemble, and promote political action, should not take this risk lightly as once an AI-powered dictatorship is introduced it might take generations to defeat it. How can we possibly overturn a dictatorship if we no longer have the ability to freely share our ideas and organize?

## CHAPTER'S KEY POINTS

- The new frontier of privacy protection is around how data is aggregated and used by algorithms.
- The ability of governments and private firms to monitor all aspects of our lives represents an existential threat to democracy.

# Cybersecurity Risk

The weaponization of ai for cyber attacks

Artificial intelligence is increasingly central to cybersecurity, not only as a defensive tool but also as an enabler of more sophisticated forms of intrusion. The same computational capabilities that strengthen detection and response can be repurposed to expand the scale and precision of attack at unprecedented speeds and levels of complexity.

Generative models can now produce highly tailored phishing communications, adjusting tone, vocabulary, and structure to align with preferences and psychological features of specific targets. Similarly, advances in synthetic media allow attackers to fabricate convincing audio or video content, increasing the likelihood that targets will respond to instructions or disclose sensitive information. These new AI-powered tools also reduce the level of technical expertise required to conduct sophisticated operations, broadening the pool of potential attackers and lowering the threshold for disruption.

A growing concern is the emergence of "synthetic authenticity," in which digital forgeries become credible enough to undermine existing verification mechanisms. When AI-generated media can convincingly replicate identities or simulate authoritative communication, institutions face challenges in maintaining trust, especially domains where verification issues can have significant consequences.

## When AI attacks ai

Many critical services, transportation, energy, healthcare, and finance, now rely on interconnected algorithmic processes that operate semi-autonomously. These interdependencies create efficiencies but also introduce new systemic vulnerabilities. As digital infrastructures increasingly incorporate AI components, cybersecurity incidents may involve interactions not only between human adversaries and automated systems but also between AI systems themselves.

AI-driven malware, for instance, can adapt dynamically to avoid detection, modifying its behavior as it encounters new defensive barriers. Conversely, defensive systems often rely on automated classification and monitoring, producing a feedback loop in which both attack and defense evolve algorithmically. This dynamic increases the speed at which incidents unfold and can diminish the ability of human operators to intervene meaningfully once an attack is underway.

## Securing the future

Addressing AI-enabled cyber risk requires a shift from reactive defense to structural resilience. Technical safeguards remain essential, but they must be accompanied by governance frameworks that recognize the distinctive characteristics of learning systems.

Traceability is a foundational requirement. Systems operating autonomously or semi-autonomously should maintain auditable records of their decision pathways, enabling post-incident analysis and supporting accountability. Similarly, critical models must be subject to oversight mechanisms that ensure their assumptions, inputs, and limitations are transparent to human supervisors.

Diversity and redundancy are equally important. Infrastructure that relies on homogeneous models risks systemic failure if a single vulnerability is exploited. Designing systems with multiple, independently developed components, each capable of checking the others, can reduce the likelihood of uniform breakdown.

Beyond technical design, the cultivation of cyber literacy plays a crucial role. Individuals and organizations benefit from understanding both the capabilities and vulnerabilities of AI systems, enabling more informed decision-making and responsible use. Awareness reduces susceptibility to deception and strengthens the broader ecosystem of digital trust.

Finally, international coordination across multiple sectors will be necessary to address the cross-border nature of AI-enabled threats. Achieving this will require continuous attention to governance, transparency, and accountability, ensuring that the expansion of machine intelligence is matched by corresponding growth in human oversight and institutional responsibility.

## CHAPTER'S KEY POINTS

- AI expands both the scale and precision of cyber-attacks, enabling tailored phishing, convincing synthetic media, and "synthetic authenticity" that undermines traditional verification systems and erodes institutional trust.
- Cybersecurity has become an interaction between automated systems, where AI-driven malware and AI-based defenses adapt to each other in real time, accelerating incidents and reducing opportunities for effective human intervention.
- Long-term security requires structural resilience, including traceability, model diversity, cyber literacy, and international coordination across different sectors to ensure that the growing autonomy of AI systems is matched by robust oversight and accountable governance.

# Social Manipulation

Machines have an unprecedented power to manipulate individuals and communities. Whether manipulation is the main goal of those who are deploying the AI model or just an unintended consequence as a means toward different goals, such as for instance driving revenue or customer satisfaction, there are countless examples of machines having a strong impact on our beliefs and behaviors both at work and in our personal lives. At work, AI is used more and more frequently to identify objectives, monitor performance, and influence behavior. For example, it is common practice among large firms to track workers in an effort to drive compliance with protocols and increase pressure to meet productivity goals. Even in our private life, AI seems to have easily taken control of many aspects of our lives such as what news we read on our phones, what music we listen to, how we work out, who we should be friends with on social media, what route we take when driving using a navigation map, and even who we should go out with while using a dating app. In a way, AI has successfully already put us in a state of very comforting techno-hypnosis in which what we see is pre-filtered and our choices pre-determined.

## Social media

The main goal of social media platforms is to maximize screen time and clicks. They are designed to show content that is most likely to increase the probability of users looking at it or clicking on it, as those are the activities that generate revenue. From the platform's perspective, the nature of the content is mostly irrelevant. Whether it is about a family event, a funny video, or a political message, all that matters is that it will create some type of reaction in the form of screen time, re-sharing content with other users, etc. The algorithms analyze user's behavior to identify what content will most likely grab a specific user's attention. For example, if you are into cats or perhaps architecture, the algorithm will learn your preferences and take every opportunity to show content you will find interesting. This apparently harmless strategy has introduced significant risk in our society and the foundation of democratic institutions.

One risk is the creation of echo chambers: groups with specific interests or beliefs are intentionally exposed to content that is likely to appeal to them rather than just a mix of different content. So, if you are into cats, the algorithm will quickly pick that up and ensure content includes cats. This becomes particularly serious when such interests have political connotations, because the AI model, in its effort to maximize users' engagement, will ensure that they will be exposed to only very similar content rather than content providing alternative points of view. For example, if you are deemed by the algorithm to be likely to believe that planet Earth is flat or that vaccines contain microchips just because you have looked at videos popular among those who like content supporting such theories, then you will be more likely to see material proposing such views. The social media algorithm is not designed to push for one agenda, but to push for any agenda that is most likely to increase your face time and therefore its advertising revenue. The fact that users are manipulated in the process or that the platform is used by criminals or extremists is seen as an irrelevant detail. All that matters is that users watch or click on the content presented to them. This is particularly troublesome considering that an increasing amount of people get their news from social media and use that information to decide how to vote, making the echo chamber effect a perfect and dangerous propaganda tool. For example, a recent paper on the 2016 US elections and the vote on Brexit found direct evidence that bots contributed to the echo chamber that fueled such a strong political polarization. As the spread of lies travels faster and more broadly than regular news, the algorithms capitalize on the outrage caused by it to increase users' engagement. So, what ends up having the most visibility is false or extreme content, leaving minimal coverage for more factual or moderate views. This effect is then exponentially multiplied by apparently innocuous tools such as re-tweets or like or share buttons, which allow users to endorse the most absurd or toxic content with no accountability and without any opportunity to challenge those views in an actual debate. Ironically, when debates do take place, one of the participants might very well be a chatbot acting like a regular user learning at each interaction what arguments work best to convince others of specific opinions.

Looking back, given the polarization caused by social media, our society would have been in better shape today if some type of control was in place to limit its abilities to fuel misinformation and extremism. This is particularly serious because it is getting to a point in which those mechanisms have effectively started to promote not just frivolous content or silly ideas, but dangerous propaganda that question the most fundamental principles of our society, such as the democratic process, the protection of human rights, and the institutions in place to protect them. While many believed that social media was going to help global communities become more connected and allow information to be more easily accessible and shared, what we have seen so far has been the opposite: rather than being an antidote to authoritarianism it is currently one of its biggest enablers.

Another way in which social media can be used to influence society is by manipulating the outcome of elections through the manipulations of the platform's users via apparently innocuous content. For instance, something as noble as reminding users to go to the voting polls, can become a tool for political manipulation if such reminders are sent to only specific groups deemed to be more likely to vote for a certain political party. Considering how close election results tend to be in mature democracies, this type of manipulation can have a huge effect on elections' outcomes. In other situations, unsuspecting voters can be directly influenced by third parties through personalized messages aimed at promoting specific agendas, such as in the well-publicized Cambridge Analytica scandal during the 2016 US presidential elections.

Another risk that comes with social media is the platforms' abilities to intentionally manipulate people's moods at any given time as it is relatively easy to push specific material aimed at influencing users' states of mind. For example, in a famous experiment Facebook was allegedly able to intentionally alter the mood of hundreds of millions of people (without informing them) by modifying the type of content displayed in their news feed. The test proved in real time with unsuspecting users that by manipulating the content you can have an enormous impact on people's lives and their wellbeing. This experiment, once disclosed, understandably caused outrage among a wide range of stakeholders. Such programs, especially given the enormous size of those platforms can end up causing serious harm in our communities, beyond just mood changes. For example, mood changing programs aimed at reinforcing sad or painful messages might cause terrible consequences among the users already depressed and considering violent actions.

Even dating sites have been shown to have incredible power influencing users' moods and sense of confidence. For example, by intentionally inflating compatibility scores shown to users about a potential match, the popular dating site OKCupid was able to alter people's perception about their match and increase the likelihood of a successful first date. While probably done with the best intentions, this type of manipulation raised serious ethical questions. Searching for relationships is something important to people, so many saw it as disrespectful and deceiving to trick users into dates with people with lower compatibility rather than those with higher compatibility.

Another substantial way social media ends up manipulating its users is by taking advantage of people's weaknesses such as a sense of insecurity and lack of self-esteem, resulting in serious psychological issues, especially among teenagers, with often tragic consequences. The rising suicide rate among teenagers has been proven to be linked to increasing pressure exercised by social media. This manipulation is resulting in serious psychological problems for an entire generation. According to a 2019 study from the US Centers for Disease Control and Prevention the death rates by suicide were stable in the decade before 2007. In the decade after 2007 it has increased dramatically and tripled among 10 to 14-year-olds. This type of negative effect on young people has been easily predictable and avoidable, but the tech companies' ethical efforts have not resulted in meaningful action to stop the harm. Many in fact now have started asking for the introduction of rules to prevent access to social media until children have passed through puberty. The answer from social media companies so far is that it is not their job to monitor who accesses their platforms and that they prefer to "empower" parents to control access instead. In other words, that's a component of the business they would rather not address.

A somewhat related issue is the fact that social media tools create a form of psychological addiction among users in which the addiction is not caused by substances but by behaviors. Many of us start our day by looking at social media and continue to do so, sometimes obsessively, dozens of times a day, especially when we are involved in activities with quantitative rewards such as the number of "likes" or views. The most troubling part is that this addiction is not a side effect, but intentionally created by design with the help of psychologists leveraging the same pattern of "reward learning" experienced by animals when seeking food.

## Web search engines

Web search engines also have the ability to easily manipulate users. Similarly to social media, web search algorithms' main goal is to show what users are most likely interested in. Results are in fact customized based on personal profiles, historical searches, and most popular clicks. This practice tends to manipulate our view of the world in multiple ways.

First, web search engines tend to promote standardization at the expense of plurality, or variety in the responses. If we search for images of a beautiful person, for example, the algorithm will tend to reinforce the mainstream view of ideal beauty with very little incentive to propose alternative views. The algorithm's objective is in fact to display first the image which users are most likely to click on, therefore reinforcing the strength of the mainstream view at each iteration. This mechanism does not only affect images, as it also affects the order by which results are displayed when looking for a specific product or service. For example, when searching for firms providing services, the algorithm will inevitably tend to give priority to large firms rather than local small ones as historically they are less popular among users.

Second, by comparing past search inquiries and popular content among users, web search engines tend to promote radicalization. For example, if you are a parent looking for info on vaccination options for your kids, the search engine might very likely show at the top of the results page sites with anti-vax narratives because those are the sites users tend to click on when self-educating on vaccines. Just as another example, when users searched "who won the popular vote" after the 2016 US presidential election in which Hillary Clinton lost the election to Donald Trump despite winning the majority of the popular votes, search engines contributed to the radicalization by giving preference in search results to popular conspiracy sites claiming that Trump won the popular vote.

This creates a true dilemma for search engine companies. Should they manipulate the results to hide questionable sites even if they are more popular among users?

## Fake news

Every time there is a hurricane in the US a photo of a shark swimming across a flooded highway dominates social media. These types of fakes might be fun if you know they are fake. The issue is that photo editing via tools like Photoshop or Generative AI, is getting so realistic that an average citizen is unable to recognize when it is not real. And thanks to AI, it can be easily deployed at a massive scale with very little cost.

Current technology allows us to create not only fake photos but also fake voices and even videos, commonly called deepfakes. One of the most famous examples is a highly realistic fake video of a speech given by former US president Barack Obama. The good thing in that specific situation is that the video was created for a good cause, raising awareness about the dangers of deep fakes, but it could have been much worse.

The risks that come with fake news, videos, photos, etc., are enormous as they directly undermine the fabric of our society, and the foundation of the democratic process such as discussing problems, sharing ideas, etc. If citizens are distracted or manipulated by fake information, or unsure whether something they see on their screens is a fact or fiction, it is hard to imagine having a healthy democracy. This is even more true if it is in conjunction with the echo chambers driven by search engines and social media because of lack of pluralism (not seeing a news story you are not deemed likely to read), which could be as damaging as fake news. One of the promises of the digital era was to allow everyone to be better informed. The power of AI has already demonstrated its ability to deliver the exact opposite.

Chatbots and Personal Robots

As AI's text and voice recognition abilities continue to evolve there are more and more situations in which humans interact directly with machines such as virtual assistant applications residing in our phones, cars, home devices, etc. The technology has improved so much that in some instances, such as when text messaging with a customer care agent, we might think we are interacting with a real person when it is in fact just a chatbot. This type of manipulation where we are led to believe we are interacting with a human when it is in fact a machine, or the other way around, can be harmful as we tend to behave differently depending on whether we believe we are interacting with people vs. machines. When interacting with machines we tend to feel less obligated to behave respectfully or politely. Experiments have shown that people display more offensive behaviors when talking to virtual assistants and bots. Mistreating virtual assistants such as Siri or Alexa is incredibly common. When Microsoft released its autonomous bot, Tay, to independently learn via interactions with social media users, the bot quickly learned outrageously offensive language, forcing Microsoft to pull the experiment after just one day.

In response to those risks, some countries have started setting up legislation introducing disclosure requirements for services supported by bots. Especially when bots are left free to interact with people, there is a high risk of AI trying to maximize the chance of getting people's attention by creating messages that end up spreading misinformation or even by capitalizing on people's emotional weaknesses. The risk of manipulation tends to be even higher when humans interact with robots. Humans tend, in fact, to sympathize with objects with shapes that resemble living organisms, especially when they are able to simulate feelings and emotion. This human trait can be easily exploited by convincing us to lower our barriers and be manipulated to behave in a certain way.

### Social credits

An emerging form of social manipulation is via social credit platforms being tested by several governments across the globe. Rather than receiving scores as buyers or sellers on specific platforms, people are receiving scores based on their behavior as citizens. In the experiment being conducted in Asia for instance, citizens are given points based on social media data or when completing honorable activities such as visiting their grandparents. Points can also be subtracted when perpetrating disreputable actions such as for instance failing to pay for public transportation. The number of points can then determine access to specific services or better pricing for products.

This is a completely new application for which there is not enough data to fully comprehend the associated risks. It is however reasonable to expect that this type of application will tend to be implemented in conjunction with highly invasive state-run surveillance programs, which will likely come with serious privacy risks and human rights violations.

## CHAPTER'S KEY POINTS

- AI has an unprecedented power to manipulate individuals and communities.
- Algorithms used by web search engines and social media tend to increase the polarization and radicalization of our society.
- A high risk of manipulation exists when humans interact with machines, especially when they are able to simulate feelings and emotion.
- Fake news and lack of plurality represent a serious risk to democracy.

# Lack of Transparency

## The Black Box problem

With machine learning, the model is able to make predictions based on rules that are created by the machine itself based on the patterns identified in the data. Those predictions are often in the form of a risk score, such as the likelihood of having a certain disease, being successful in college or at a job, or committing a crime. Those models are called black boxes because they operate in a way that is too complex for people to retrace and fully understand how the information is processed. Even the algorithms' developers themselves do not fully understand how they actually work. This opacity conflicts with key values like transparency and explainability.

The lack of transparency is largely around the numerous decisions made throughout the development of the technology, such as the type of data selected to feed a model and the design of the algorithm. It might also be related to the fact that the nature of a technology being used for a specific task (AI-based vs. other types) is not clearly disclosed to users and other stakeholders. This is particularly important considering that algorithms are often used in support of sensitive processes such as college applications or HR decisions affecting people's lives. For example, if my job application is declined even before I have a chance to have my first interview, it would be important for me to know whether the rejection was the outcome of a review made by the recruiter or by a resume scanning algorithm. However, it is increasingly more difficult for the public to identify the nature of technology supporting their transactions and interactions. It is at times difficult even to know whether we are interacting with an AI-driven chatbot or a real person. With the recent advancement seen with Generative AI, the risk of lack of transparency will touch even more areas. For example, a person reading a news article might not realize to be reading something generated by an AI model rather than a journalist. Companies have no incentive to disclose to users that they are interacting with chatbots as it has been proven that they tend to deliver better results (especially sales) when people are not aware they are dealing with AI. Transparency issues have emerged not only due to the failure of disclosing the use of AI, but also the other way around, with firms creating false impressions of delivering AI-driven capabilities. For instance, a researcher ordering food at a restaurant via a chatbot discovered that the online chat was not run by a software but by a person pretending to be an AI model. The researcher, who got suspicious about the type of answers received by the bot, was finally able to get the bot to admit to being a real person. In much more concerning cases of AI impersonation, companies routinely sell solutions presented as AI-driven that are actually manually performed by workers from low-income countries. For example, an AI solution sold to stores to detect theft in real time with the use of live smart cameras was recently exposed as actually being exclusively supported by manual work sourced from low-income African countries.

The lack of explainability is a narrower concept and it has to do with the way the information is processed by the model and how the input is manipulated and turned into the model's output.

Sometimes the lack of explainability is due to the fact that the logic used by the model is protected for business-related interests by the firm that developed the algorithm. So, the lack of explainability is just due to a firm's decision not to disclose their trade secrets. The secrecy is understandable, but it is also a risk as it makes it very difficult to detect mistakes and fix them. In the HR example above, if the rejection is the outcome of a determination made by the algorithm, nobody would ever know, even the recruiter, why a candidate's resume was discarded.

Some other times it might be that the model is so complex that its explanation is not understandable unless you are an AI expert, so some level of explainability exists but only within those with in-depth training in the field.

In other situations the lack of explainability is due the insurmountable complexity of the model which makes it impossible for humans to interpret it as there are too many complex layers, and no person can possibly determine what caused the model to provide a certain answer. There is therefore no explanation of how the model gets to a certain answer. This lack of explainability is one of the most polarizing topics in the AI community. For some experts, it is a highly concerning risk as it creates situations in which decisions are made without ever having a full understanding of the drivers. As a result, the people affected by it, such as job candidates or recruiters, do not have any way to know why a machine came up with a specific recommendation. For others it is less of a concern as a complete explanation of every component of a phenomenon is rarely needed anyway as a general explanation is often sufficient. For example, if the highway has slower traffic than usual and we want to know why, it might be sufficient to simply know that it is due to an accident, without necessarily getting into all the situations that caused every single vehicle to be on the road.

## Tradeoff between explainability and accuracy

While there are benefits for deploying black boxes, their opacity poses important ethical questions, especially when algorithms are used in so many key areas of our lives such as public health or criminal justice. Should defendants always have the right to know if the sentencing is developed based on a black box? Should they be given the option to request a trial fully based on humans' decisions only? How can we ensure due process in a trial if no one can understand how exactly the black box that a judge is relying on came up with a certain sentencing recommendation?

Given that the lack of explainability is going to increase as more complex models are developed, the real dilemma is around the tradeoff between explainability and accuracy. Are we OK as a society with well-performing algorithms providing very accurate predictions even when there are no clear explanations? For example, algorithms have proven to be incredibly accurate in the detection of pulmonary hypertension. Those predictions are based on millions of correlations that no one can really understand, but they are consistently accurate. Is it ethical to not use that model and therefore choose to have higher rates of misdiagnosis just because of lack of explainability? Additionally, is explainability always a must have? If we think about our criminal justice system, don't we have countless situations, such as jury trials, in which inconsistent and unexplainable outcomes are still accepted? Even in the medical field we have procedures that we really do not understand but still use because of their benefits: from general anesthesia to electroconvulsive therapy, there are plenty of situations in which we seem OK with using techniques without having clear explanations of how they work.

## Addressing the models' opacity

The opacity problem will continue to grow as models reach higher levels of complexity. This represents an emerging risk, as we can expect to understand less and less about how machines operate. This lack of transparency could potentially be used to avoid responsibility for the deployment of unsafe or biased models.

Some researchers have proposed new techniques that can help minimize such risks by analyzing the determining factors of certain outcomes. For example, while it might not be possible to explain every step taken by the algorithm, if you run models tweaking different inputs, it is possible to identify what factors influenced the outcome the most. Going back to the HR example, if a resume is rejected by a black box, using this technique it is possible to at least identify the leading disqualifiers by fine-tuning different components of the resumes (years of experience, keywords, etc.), and see what changes would have resulted in an approval.

In situations in which algorithms are used for image recognition tasks, the opacity can be also at least partially addressed by creating segmentation in the images used by the AI model. For example, if you look at the software used for cancer diagnosis by scanning x-rays, there are techniques that allow you to design the model so that the image is divided in quadrants and instruct the algorithm to indicate which quadrant is triggering a positive response. While it does not allow you to know what exactly triggered a positive diagnosis, you can at least know which specific image's segment was relevant in that outcome.

## CHAPTER'S KEY POINTS

- Models using machine learning process data in a way that is often too complex for people to retrace and fully understand.
- Explainability might turn out to be something we have to give up in exchange for the benefits we get from black boxes.

# Quality and Safety Issues

The risks related to AI quality and safety are vast as they touch a variety of issues from product safety to value misalignment and long-term existential risk. This chapter will focus specifically on safety risk related to product reliability, robustness, and security. The other components will be covered later in the book.

## Major failures

Given the rapidly evolving nature of new technologies, companies tend to underestimate the risk. Even if AI is only used for very narrow applications, we have already seen a number of failures associated with the deployment of AI caused by the lack of proper controls to minimize the risk of harm due to defects, misuse, or unintended consequences.

Just by looking at a company such as Amazon, there have been countless situations in which failures quickly translated into negative press. In one story, it came to light that a faulty book recommendation algorithm was showing publications developed to train terrorists among the recommended books on its online bookstore. In another story, Amazon's AI-driven pricing model was found to price individual books for millions of dollars due to an unintended redundancy in its data set. In another instance, it was reported that Amazon's Alexa mistakenly played an audiobook with adult content to a little kid who was trying to instruct Alexa to play a kid's song. In another incident, it was reported that household Alexa devices tried to order dollhouses on Amazon when they heard a host on the TV news say something like, "Alexa ordered me a doll house". In another widely publicized story, Amazon's Alexa recorded a private conversation and then accidentally sent the recording to a random person on the owner's contact list. Amazon is not the only major tech company to experience such epic failures. For example, in 2016 due to an error in the database of Google Maps, an entire home was demolished by mistake as the demolition crew relied on the map to identify the house to be demolished. If the most successful companies in the world are unable to safely deploy AI for relatively simple activities such as providing book recommendations or online ordering, are we sure we are ready to use AI in much more complex and unpredictable environments?

The flow of news reporting outrageous failures does not seem to ever stop. In 2012 the Knight Capital's AI-driven trading system, which was given the power to freely invest in the stock market, lost $10M a minute for 44 minutes until someone turned it off wiping out $440 million of the firm's capital . In another story, shares of Warren Buffett's Hathaway climbed dramatically after the algorithm misinterpreted a news story about the release of a movie with actress Anne Hathaway. In 2016, a 300-pound security robot went haywire and ran over a one-year-old boy at a California mall. In 2022 a robot built to play chess grabbed and broke a finger of a seven-year-old kid. Also in 2016, a Tesla self-driving car was involved in a fatal accident at an intersection, as it allegedly failed to recognize a truck.

A more tragic example is the famous case of the Boeing 737 Max. Due to faulty sensors combined with overly complicated override protocols for the automated flight control system, two crashes caused the loss of hundreds of lives. In 2023, relying on facial recognition software for the identification of a suspect in a theft case, Police in Louisiana wrongfully jailed for a week an innocent man that not only did not have anything to do with the crime, but also did not even match the physical description of the suspect, and had never visited the state in which the crime was convicted.

Even chatbots have proven to be able to cause severe harm due to their unreliability and tendency to create false information. For example, in answering a question around scholars who had sexually harassed someone, in 2023 ChatGPT falsely listed a real law professor among people who had been accused of sexual harassment, citing news articles that never existed. In another example, in June 2023, a New York-based law firm was fined after a judge discovered that a court brief submitted by the lawyers created with ChatGPT included citations of fake legal cases.

## Safety standards

What level of safety should be expected for AI-driven tools? The good news is that this is not the first time humanity has needed to apply risk management tools and there is a well-established methodology we can use as a starting point to assess the tolerance level for the inherent risk of a new solution. The same way the microwave oven we have at home has safety requirements so that it automatically shuts off as soon as we open the door, we need to ensure we have proper controls to reduce the risk of harm from AI models.

First, we should start by acknowledging that safety cannot be an optional attribute of AI. All AI should be designed with the objective of being safe. A civil engineer says, “I design bridges”, not “I design bridges that don’t collapse”. Similarly, software developers should consider safety as an implied imperative, and not like it is an option. Some quality issues are going to be easier to identify than others. For example, if we are building a model that will be used to recognize street signs or to turn speech into text, it is relatively easy for humans to identify an error and take action. In other areas it is much more difficult for humans to identify gaps in the model as no sufficient data exists. For example, if a machine is asked to predict criminal recidivism, the risk of terrorist actions, or the quality of future job performance, it is impossible to truly test the model to ensure accuracy as there is no way to foresee what the future will look like. In such scenarios we might be better off without an AI-driven model as it would be impossible to audit the reliability of the algorithms' predictions.

Second, we need to define what an acceptable level of safety would be. It is easy to say that it needs to be 100% safe, but is 100% reasonable or even achievable? Engineers have safety standards that change depending on the actual risks. For example, if the bridge is for a seismically active area, there are standards in place based on the expected frequency and severity of earthquakes. As we do when we design a bridge, we need to set safety standards that are specific to how and where the technology might be used: What is the intended use? In what context? How long will it last? What could go wrong? Who might be affected? What controls are needed? For instance, our tolerance for network vulnerabilities should be significantly lowered if we are building an AI-driven smart implant to be used for medical purposes. Can you imagine the consequences of having an autonomous liver that gets hacked? Similarly, before we flood our streets with self-driving vehicles, we need to agree on the safety standards we expect for them. If we think about a self-driving taxi, we need to ensure it is going to be on average at least as safe as the taxi driven by a person across all types of scenarios including different amounts of traffic, weather conditions, and times of day, and the potential presence of distracted pedestrians, etc. It also needs to have sufficient cybersecurity safeguards as the consequences of a successful hacker's attack might be disastrous.

Third, similarly to what we do for other technologies, we need to create standards around investigations of accidents caused by intelligent machines so that we can learn how to prevent them in the future. All of this does not exist yet.

## The Precautionary Principle

The higher the uncertainty, the more caution should be used. As appropriately captured in the "precautionary principle" developed centuries ago in the civil engineering sector, we should not wait for scientific certainty on the risks in order to take the right precautions. The fact that we lack sufficient knowledge to assess the risks of AI should be a reason for caution rather than blindly jumping into it. The worst thing we can do is to continue to build even more powerful machines without first taking the time to understand how they work, to ensure they do not cause harm.

## CHAPTER'S KEY POINTS

- While AI is only used for very narrow applications, we have already seen a number of failures associated with the deployment of AI.
- Safety standards need to be developed for each application on AI based on how, when, and where the technology might be used.

# Value Misalignment

The genie is out of the bottle

Many cultures have stories involving some type of a genie with amazing powers that can grant you anything you wish. In Greek mythology, there is the famous story of King Midas. When granted a wish from one of the Gods, he chose to have everything he touched turned into gold, making him the richest man. After the initial excitement, he realized that he could no longer do anything, even eat, as everything instantly turned into gold.

One of the biggest risks with AI is exactly the same: if there is any ambiguity about what we ask AI to do, it will be used against us because we will probably get what we asked for but not what we wanted. While AI has the amazing power to solve complex problems, it does not possess the power to understand what we want. When we give objectives, there are always a number of values or assumptions or implied objectives that are in our head and need to be spelled out if we want the machine to behave in a way that is in line with our goals and values. For example, if you ask a machine to assemble pieces of a humanoid robot (arms, legs, etc.) so that the robot can go from point A to point B, the machine will find the best way to do it: rather that doing what you want it to do (build a machine that has legs and moves around walking) it might very likely assemble the robot to look like a long stick made of parts (arms, legs, etc.) that line up from point A to point B. Machines only know what we tell them. They are not able to guess what solution we might be thinking about to reach the objective. Similarly, if I asked a robot baby-sitter to feed the kids, I would have to be very explicit about what is okay and what is not okay to feed them and how, and that the food should be offered, not force fed, etc. Just as a real-life example, a robot being trained to play soccer and given the goal of touching the ball as many times as possible, quickly figured out that the best way to do so was to stand next to the ball and rapidly vibrate.

Thanks to Hollywood movies, when we think about things that might go wrong with AI, we tend to think about robots turning evil and shooting people left and right. Despite what we see in movies or read in Sci-Fi books, nobody should be worried about machines turning evil. Machines do not have emotions or personal agendas. They do not wake up one day with malevolent plans. The risk originates from the way they are programmed to achieve goals. This is perfectly represented in Stanley Kubrick's movie 2001: A Space Odyssey. The machine controlling the spaceship tries to defend itself against any of the astronauts' attempts to switch it off not because it becomes evil, but because it knows that once turned off it will not be able to complete its task. The better AI becomes at accomplishing goals, the more important it will be to ensure that those goals are clear and aligned with ours.

At least for the foreseeable future, the most likely scenarios in which machines hurt humans are when we intentionally or unintentionally tell them to do so due to our own unethical behavior or incompetence. Intentionally: when we deploy them to perform activities that cause harm to people, as with lethal autonomous weapons. Unintentionally: when we deploy them to complete a task without removing ambiguity from the stated objectives. For example, we might ask a machine to help maximize the chances of the survival of the human species and, in response, the machine seeks the elimination of a vast portion of the world's population as a way to best reach its goal. This risk of such misalignment becomes particularly high when meeting an objective requires the AI model to solve for a number of sub-objectives for which there is little or no human guidance. For instance, if we ask an AI-powered personal robot to go to the store to buy milk, the machine will necessarily identify a number of objectives (getting to the store, finding the right aisle, etc.) that will need to be met as part of the strategy to achieve the main goal. Such objectives will come with countless scenarios in which the machine might deviate from what we envisioned. For example, it might end up hurting people along the way, damaging property, failing to wait for its turn in the check-out line, etc.

An even riskier scenario is one in which humans might decide to let the machine choose a goal. At that point the model would simply start calculating what an AI machine would most likely want by searching for clues in the available data (documents, movies, etc.). What if the AI model found its answer based on catastrophic Sci-Fi stories where machines enslave all humanity and identifies that scenario as its own objective? Fortunately, this is all very unlikely, but it would be irresponsible not to have controls in place. As we have seen, people tend to get easily attached to anthropomorphic machines. Even with fairly unsightly machines such as Roomba vacuum cleaners, people often feel bad when those machines hit a wall or get stuck under the couch. Sooner or later there is going to be someone that will have such strong enough feelings toward a machine, that they will give it carte blanche, based on guilt or perhaps love. It could be like a father feeling guilty for forgetting the birthday of his daughter, taking her to the store and asking her to choose one present. If that ever happened with AI, that's when things could really get out of control. Once again, the most concerning aspect of AI is not AI's malevolence, but human incompetence.

## Eliminating ambiguity

Since it is so difficult to tell machines what we really want, how do we ensure that machines operate in line with our values? To eliminate the risk of misalignment between our goals and the objectives given to machines, we need to avoid any risk of ambiguity in what we ask AI to do. To do so, we need to first understand what we really want them to do, then be extremely clear about how we convey those objectives, including the implied ones. So, a big part of the solution is to learn how to better communicate with AI.

Some believe that we cannot really address the risk unless we preprogram every single step from beginning to end, addressing all types of foreseeable situations. This, however, not only would be very time consuming but would still come with a high level of risk given the endless scenarios a machine might face in the real world. For example, when programming a self-driving car there are so many variables and types of behaviors to consider (example: kids on scooters riding on the wrong side of the road) that it would be practically impossible to account for all of them.

Others believe that machines themselves should identify all the specific rules to behave correctly, by only giving them a high-level goal. Given the rule, the machine will figure out the most appropriate behavior through trial and error. For example, if it is an algorithm running a self-driving car, you will teach the machine the right behavior by giving points for each mile driven and deducting points for each person hurt. After a while, the algorithm will learn how to drive long distances without hitting pedestrians. This is generally referred to as reinforced learning.

Other experts, such as Stuart Russell, are proponents of imitation learning techniques where machines learn values by observing human actions. Rather than being told what to do and how, they are tasked with figuring out the objective just by observing human behavior. In the self-driving car example, the machine is not told to avoid hitting pedestrians, it is only shown how to drive well. The machine will eventually come up with that rule either independently or with a set of exchanges to reinforce the best option. This is generally referred to as inverted reinforced learning. Rather than giving machines certain objectives, you intentionally leave the machine constantly questioning what it knows about what we want and humbly looking for clues. A modern version of Socrates' famous statement "The only true wisdom is in knowing you know nothing" strategically injected into AI. The most obvious challenge of this approach, knowing how inconsistent our actions are if compared to our own beliefs, is ensuring that the machine trying to extrapolate our values only sees behaviors that are consistent with them.

Finally, other experts, such as cognitive scientist Gary Marcus, have suggested that the solution might ultimately end up being some type of hybrid combination between the flexibility provided by Machine Learning and the trustworthiness of the good old-fashioned rule-based AI.

## CHAPTER'S KEY POINTS

- Machines will always do what we ask them to do, but that might be different from what we want them to do.
- To eliminate the risk of misalignment between our goals and the objectives given to machines, we need to avoid any type of ambiguity in what we ask AI to do.

# Inadequate Human Oversight

The emerging importance of Human Oversight

As AI systems are deployed in increasingly complex contexts, the concept of the "human in the loop" (HITL) continues to gain increased relevance. Complexity, whether in the form of high-stakes applications, multi-step tasks, or ambiguous decision environments, heightens the need for human judgment, particularly in scenarios where absolute accuracy or contextual sensitivity is required. Human oversight is not merely about ensuring that someone is somewhat involved: some participation by users, managers, compliance officers, executives, and boards will inevitably always be at least partially present. Rather, it is about having specific individuals with clear accountability for the accuracy and appropriateness of the execution of a specific process, with the ability to evaluate and, when necessary, correct the system's outputs.

Human oversight may be necessary for several reasons. In some cases, it provides a corrective mechanism: humans identify and address errors that the system cannot detect. In other cases, oversight serves as a way to detect and address situations involving competing priorities, ambiguity, or situational nuance requiring a judgment call based on values and contextual considerations. There are also domains where delegation feels ethically inappropriate such as decisions involving life, safety, or irreversible harm.

Oversight itself occurs at multiple levels. It is important to distinguish between evaluating the performance of a system as a whole and reviewing individual outputs. A program manager assessing the accuracy of a code-generation platform exercises oversight at the system level to ensure the platform itself is performing as well as expected. An individual developer reviewing each suggested line of code provides oversight at the output level to validate the quality of the AI-generated code. Both functions matter, but they address different questions: whether the system is behaving reliably, and whether a given output is appropriate for a particular context.

Recognizing these distinctions is essential for designing effective governance frameworks. HITL is not a single mechanism but a constellation of human roles, each embedded within a broader organizational structure. Understanding how these layers interact is a prerequisite for determining where oversight is most needed and how it should operate.

## Humans in the Loop

Human oversight is often framed as a simple yes-or-no safeguard, but this characterization obscures its underlying complexity. Oversight varies widely in scope. Treating it as a binary condition risks overlooking its fluidity and nuances. Two dimensions are particularly relevant: the nature of oversight and its timing.

The nature of the oversight function can take several forms. In its most stringent configuration, HITL, the human acts as the approver of each output, with the process advancing only upon explicit confirmation. This model is particularly appropriate when tasks involve high risk, significant regulatory implications, or irreversible consequences. A less intensive model, often referred to as "human on the loop" (HOTL), positions the human as a watcher, monitoring system behavior, and intervening only when anomalies arise, retaining authority to pause or override system actions. This form balances autonomy with control but depends highly on the watcher's attentiveness and expertise.

Timing is equally important. As in the HITL and HOTL examples above, oversight can occur in real time, where humans evaluate outputs before they take effect. This provides strong safeguards but may introduce operational delays and cognitive load. Alternatively, oversight can be retrospective. Quality assurance processes, periodic audits, and post-hoc evaluations fall into this category. These mechanisms are essential for detecting systemic issues, bias patterns, or long-term drift, even though they cannot prevent errors in the moment.

Across all scenarios, HITL requires clarity on AI's autonomy and guardrails, expectations and decision-making authority of the people assigned to monitoring activities, and adequate institutional support. Reviewers must have access to relevant information, sufficient time to evaluate outputs, and formal authority to intervene. Ambiguous mandates risk creating situations in which oversight exists in theory but is ineffective in practice.

## Risks to effective oversight

The effectiveness of human oversight might be compromised by several scenarios.

One scenario of risk is when human oversight is expected but not adequately resourced. Effective review requires sufficient time to examine outputs with care, access to the relevant data, and the expertise necessary to evaluate whether an AI system's outputs align with technical, legal, or ethical standards. When any of these elements is lacking, reviewers may overlook errors or other harmful outcomes.

Another scenario has to do with speed. Humans are often too slow to provide real-time adjustments, especially when decisions occur in milliseconds or involve dozens of intermediate states invisible to the reviewer.

Another frequently discussed risk is complacency and overreliance, a situation in which humans assume that the system is functioning correctly and reduce their vigilance accordingly. When systems perform reliably most of the time, humans may implicitly trust them by default. Over time, oversight may become a form of auto-approving in which reviewers routinely confirm outputs without fully evaluating them. Empirical research on automation bias demonstrates that humans tend to defer to automated recommendations, particularly under time pressure or informational overload. This tendency, sometimes described as "automation complacency", is well documented in aviation, medical diagnostics, and financial risk scoring, where human oversight easily deteriorates to rubber-stamping.

A deeper and more subtle concern is the erosion of human fluency. Over extended periods, reliance on AI systems may diminish the ability to independently perform the tasks the humans in the loop are assigned to review. If oversight requires domain expertise that humans gradually lose, the oversight mechanism itself becomes fragile. This dynamic is particularly concerning in high-stakes fields where skill degradation may be difficult to reverse.

These risks underscore a broader principle: HITL's effectiveness depends on the design of the oversight role, the competence of the individuals involved, and the organizational conditions supporting them. Ensuring that human judgment remains a meaningful counterbalance to AI's autonomy requires continuous investment in training, periodic recalibration of oversight responsibilities, and recognition that human involvement is not a static safeguard but an evolving component of responsible AI governance.

## CHAPTER'S KEY POINTS

- Human oversight becomes more essential as AI tasks grow in complexity, especially in contexts requiring nuanced judgment, contextual awareness, or ethically sensitive decision-making.
- Human oversight operates at multiple levels, including system-level evaluation (overall performance) and output-level review (individual outputs), each addressing different governance needs.
- Effective oversight depends on clarity of role and timing, distinguishing between real-time approval and backward-looking quality assurance.
- Human-in-the-loop approaches carry their own risks, including lack of resources, over-reliance, and erosion of human expertise, underscoring the need for sustained training and institutional support.

# Labor Displacement

## A continuation of the automation journey

Few topics in the AI debate generate as much attention, anxiety, and disagreement as its impact on work. Every major technological shift has reshaped labor, disrupted existing professions, and forced societies to rethink how people earn a living. Since the dawn of agriculture, humanity has excelled at inventing tools that maximize output while minimizing physical labor. With AI, we have begun building something different: systems capable of interpreting information, producing language, making predictions, coordinating workflows, classifying data, and generating content. As long as machines' profitability is higher than the cost of designing, building, and running them (mostly hardware and electricity), or there are other benefits related to performance and efficiencies, there will be strong incentives for firms to use them whenever possible. What does this mean for the future of work? How will jobs be reshaped? How will labor markets evolve? Which risks demand the most urgent attention as AI becomes more deeply embedded in our lives?

## Impact on Current Jobs

AI is already reshaping work in concrete ways, improving efficiency and safety in some contexts while introducing new risks that fundamentally alter how jobs are performed and governed.

In some areas, this change has been clearly positive. AI systems have freed workers from repetitive and time-consuming tasks: sorting emails, scheduling meetings, drafting standard documents, summarizing long reports, or searching through large datasets. In creative fields, AI tools can generate images, text, and music, accelerating early stages of production and allowing professionals to produce more quickly. In these contexts, AI functions as a formidable augmenting tool. In certain industries, AI has also helped improve safety. In transportation, driver-assistance systems use sensors and computer vision to detect obstacles, pedestrians, and hazardous conditions, helping reduce accidents. In manufacturing and energy, AI-enabled monitoring can detect equipment anomalies early, preventing failures that could cause injury or significant damage.

AI has introduced serious downsides as well. In some workplaces, AI has made jobs more dangerous rather than safer. Studies have shown that warehouse workers operating alongside large autonomous robots experience higher rates of injury than workers in facilities without them. Even more significant is the expansion of workplace surveillance. AI-enabled monitoring systems now track workers' movements, productivity, and behavior in real time. This trend has been especially pronounced among blue-collar workers in logistics, warehousing, and delivery services, where algorithms monitor walking speed, driving behavior, route deviations, and task completion rates. Employee surveillance is now spreading to white-collar work as well. Large organizations increasingly track keyboard strokes, mouse movements, meeting participation, facial expressions during video calls, and even vocal tone to infer engagement or emotional state. Meetings may be recorded, transcribed, and analyzed by AI assistants. AI is also playing a growing role in decisions that shape careers. Algorithmic systems screen resumes, rank candidates, evaluate performance, and flag "high-risk" employees. These systems often lack transparency and can perpetuate biases. Workers may have limited ability to contest outcomes, because the decision process is hidden behind proprietary models and complex statistical reasoning. At the same time, AI increasingly directs work itself. In many settings, workers receive instructions not from human supervisors but from algorithms that dynamically allocate tasks based on real-time demand. Car ride platforms pioneered this model: drivers follow the app's instructions, accept jobs assigned by algorithms, and are evaluated by automated systems. But similar approaches are now being adopted in factories, warehouses, and service industries. Work is increasingly coordinated by machines rather than people.

## Impact on employment market

As seen in every major technological revolution, AI will follow a similar trajectory. Some jobs will disappear, some will be offshored, some will significantly change, some will only be minimally impacted, and new ones will emerge.

In some cases, it is reasonable to expect that entire professions will vanish. Just as telephone switchboard operators or elevator attendants were eliminated by earlier waves of automation, some occupations will cease to exist altogether. This will not be limited to blue-collar work, nor confined to white-collar roles. Jobs across a wide spectrum, from traders, analysts, customer service representatives, warehouse workers, food service employees, are vulnerable in different ways. Even in grocery stores, checkout associates have already been widely replaced by self-checkout systems, where a single worker can oversee many machines. Even that role may be temporary. There is no obvious technical barrier preventing supervision itself from being automated over time. Cameras, sensors, anomaly detection, and predictive maintenance can reduce the need for human oversight. Displacement in one occupation can also trigger cascading effects across local economies. If long-haul truck drivers are eventually replaced by autonomous vehicles, the impact will extend far beyond drivers themselves. Rest stops, roadside diners, motels, repair shops, and other businesses that depend on that workforce would also be affected.

Many attempts have been made to quantify how much work AI will automate. Some studies produce striking figures. According to a study from Oxford a few years ago, 47% of US jobs will be automated. When a similar analysis was applied to China and India, it was 69% and 75%. According to the International Federation of Robotics, the automation of production is exponentially accelerating around the world with 126 robots per 10,000 employees in 2021 vs 16 in 2015. Much of the forecasting, however, remains closer to educated speculation than to robust prediction as they rely on simplified (if not unrealistic) assumptions, such as no future changes in technology or regulations, disregarding the fact that societies respond, regulations change, and new technology emerges.

In many cases, AI will not eliminate entire jobs but automate components of them. This will include professions typically considered immune to automation such as attorneys or doctors, especially with regard to specific work involving repetitive and predictable tasks, such as day-to-day administrative tasks like scheduling, sorting emails, researching documents, creating documents, etc. As AI takes over transactional or data heavy tasks, the work performed by humans shifts toward more judgment-heavy or ethically charged activities. In the radiology field, for example, AI can scan medical images faster and sometimes more accurately than humans, yet radiologists remain in demand because diagnosis involves more than detection: it involves communication, empathy, and deep knowledge across multiple medical domains that AI cannot fully deliver. Similarly, most customer inquiries submitted to contact centers will be handled by AI directly, leaving to humans only the most complex cases. If this continues to be true, the functions and professions most at risk are not necessarily those whose tasks can be automated, but those who fail to adopt tools that change their field. In many contexts, the competition is not "human versus AI," but "human with AI versus human without AI."

Another dimension of displacement is not elimination but relocation where jobs are moved to other countries. A striking example is the labor required to train and validate AI systems. Content moderation, image labeling, audio categorization, and the review of harmful text are frequently performed by workers in low-wage countries. This is ethically significant because these jobs are often precarious and psychologically difficult. They require sustained exposure to disturbing material (violent content, hate speech, and sexually explicit text, etc.) so that models can learn to detect and filter it. Offshoring this work is driven not only by cost reduction but also by the desire to limit legal liability and reputational risk by making it less visible. Although relatively unknown to the public, this practice is widespread. In a widely reported case, it emerged that OpenAI's ChatGPT relied on Kenyan workers paid less than two dollars an hour to review and label text containing violence, hate speech, and sexual abuse. This example highlights another key point: while automation is sometimes framed as inevitable job loss in the name of progress, sometime it is just a way to relocate jobs to cheaper workers elsewhere.

Not all jobs disappear or transform radically. There are also professions that AI struggles to replace, particularly those involving complex critical thinking, social relationships, ethical judgment, ambiguity, and operation in unpredictable environments. Teachers, social workers, caregivers, moderators, and many skilled trades fall into this category. In these fields, AI can support planning, documentation, and pattern recognition, but it does not reliably replicate the relational, contextual, and situational dimensions that define the work itself. This does not mean these jobs are immune to change. It means that automation may function more as augmentation than replacement. AI can assist teachers with lesson planning, but it does not replace human relationships. AI can support caregivers with scheduling or monitoring, but it does not replace embodied care. Even when parts of a job can be automated with AI, the job itself may remain human-led. Consider the case of self-driving trucks. Even if autonomous driving becomes reliable, firms may still want a person on board to handle ancillary tasks: checking the vehicle, managing unexpected physical obstacles, handling manual loading, dealing with regulatory inspection, making quick safety judgments in ambiguous contexts, or responding when the system fails.

At the same time, technological change continues to generate new roles. Data scientists, AI ethicists, prompt engineers, and machine learning operations specialists are only some examples.

### Will the economic pie continue to expand?

Given that AI will affect different jobs in different ways, the central question is whether automation will eventually reduce demand for human labor faster than societies can generate new types of jobs.

These fears have accompanied every major technological revolution. Yet since the Industrial Revolution, no technology has produced permanent mass unemployment. Instead, jobs have shifted across sectors, new industries have emerged, and living standards have overall improved. Before 1800, up to 90% of people worked in agriculture. Today, agricultural employment in advanced economies has fallen to the low single digits. And yet total employment has grown. The jobs did not disappear, they moved. Labor shifted into services, healthcare, education, technology, management, and many other areas that expanded as societies became wealthier. Will the same dynamics continue to be true with AI?

Many believe that there will never be a lack of work because human labor will remain in demand for at least two reasons.

The first reason is that new jobs will continue to emerge. Automation reduces the cost of goods and services. Lower costs often lead to higher demand. Higher demand requires more production, more coordination, and more service, much of which remains human-intensive. Automating parts of the transportation sector, for instance, could make travel cheaper. Cheaper travel tends to increase travel volume, which increases demand in adjacent industries: hospitality, infrastructure maintenance, customer service, experience design, regulation, and many forms of local labor that are difficult to fully automate. Similar dynamics occur in healthcare, education, entertainment, and many service sectors. Technology shifts the boundary of what is affordable, and affordability expands markets. While public narratives often focus on AI displacing workers, a less visible but equally important effect is that automation expands the overall economic "pie," creating new areas in which human labor becomes valuable. By shifting the boundary of what is economically feasible, technology enables new kinds of work that previously did not make economic sense. Even as AI takes over more tasks and occupations, overall employment can continue to rise because productivity gains unlock new activities and markets. In this dynamic, the amount of work performed by machines may expand, yet demand for human labor can grow at the same time because the economic "pie" itself becomes larger. The same mechanism operates in everyday life. When time is limited, what we choose to work on depends on the tools available to us. If I am dedicated to spending an hour cleaning up dishes and taking care of my yard, the decision around what to work on will vary depending on what technology is available. If I do not have a dishwasher, I will spend most of my time washing dishes. If I have a dishwasher, I will spend most of my time mowing the lawn. If I have an autonomous lawn mower, I will spend most of the time removing the weeds, pruning the bushes, or improving other aspects of the yard.

Second, it is reasonable to expect that AI will lead to technological breakthroughs that will also create new markets. When cellphones became a consumer good, they quickly became indispensable, and an entire economy formed around them: apps, accessories, mobile payments, customer support, infrastructure, cybersecurity, digital advertising, and countless ancillary services. The key mechanism is not that needs are infinite in a literal sense, but that preferences evolve. People adapt to new capabilities and reorganize their lives around them. Thorstein Veblen captured this dynamic with the phrase "invention is the mother of necessity." The point is that inventions reshape what societies consider normal, convenient, and acceptable. Once something becomes possible, it often becomes expected. That expectation generates economic activity. Employment data seems to confirm those trends. Historical employment data supports this dynamic. In the early twentieth century, roughly two-thirds of jobs were concentrated in agriculture and manufacturing. Today that share is closer to 15 percent, yet the total number of jobs has increased dramatically. In fact, the number of available jobs today is roughly six times higher than in 1900.

To be clear, even if this view proves correct, it does not imply a future of full employment. Rather, it suggests that if AI-driven unemployment emerges, it may stem less from a lack of work to be done than from a mismatch between available jobs and available workers. People may lack the skills, geographic mobility, or institutional support required to move into roles that become newly valuable. In such a scenario, there may be many open positions, but very few workers equipped to fill them. For example, is it realistic to expect a taxi driver in his fifties to transition into a cybersecurity engineer? Even if technically possible, how often would that actually occur? Who would pay for the training, and who would support the household during the transition? These questions underscore, once again, the central role of proactive public policy. Without substantial investment in reskilling, income support, and transition assistance, it is unlikely that labor markets alone will be able to absorb the disruption.

### What if job creation cannot keep up?

Others have a less optimistic view, arguing that AI will eventually automate such a broad range of tasks that job creation will not keep pace.

One driver might be the fact that AI's output reaches such a high level of quality and efficiency that humans can no longer compete against them. Machines can be cheaper, faster, and more predictable. They do not quit. They do not unionize. They do not demand wages. As their capabilities improve, displacement could be broader and more simultaneous than ever before. The classic analogy is horses: for centuries, horses were central to transportation and logistics, and then automobiles displaced them almost completely. Humans could face a similar fate in many occupations if AI becomes good enough at a sufficiently wide range of tasks.

Another driver might be demand saturation. Even if automation reduces costs, demand does not expand without limit. At some point, for example, cheaper goods do not translate into proportionally higher consumption. When saturation is reached, productivity gains translate into reduced labor demand rather than expansion. There is a maximum amount of fruit a person will buy regardless of price.

If AI were to take over most jobs without creating enough new roles, the implications would be profound. The scenarios range from highly optimistic to deeply negative.

One optimistic view is that AI could free humans from the need to work. Automation would produce abundance. Machines would grow food, build houses, and provide services. Humans would spend more time with family, engage in creative pursuits, volunteer in their communities, and live better lives. Rather than a sign of poverty, unemployment would become an example of high standards of living and today's factories and other workplaces would be seen as a nightmare. In this scenario, rather than worrying about AI taking over our jobs, our priority should rather be to figure out individually and collectively what type of activities we want to retain for ourselves rather than delegating to machines. Many big players of the AI world do believe that this is where we are headed. For example, when Elon Musk presented in 2022 the latest version of his humanoid robot, Optimus, he stated that its development "means a future of abundance, a future where there is no poverty, where people, you can have whatever you want, in terms of products and services. It really is a fundamental transformation of civilization as we know it." In order for this to become a reality, we certainly need a high level of maturity and cooperation among humans because we would have to agree to some type of mechanism to redistribute the benefit of the increase of productivity delivered by machines. It would also need proactive governance to influence the direction of automation so that it targets the tasks we most want to eliminate and not the ones we enjoy. We imagined a future in which machines would handle plumbing, cleaning, and other physically demanding or repetitive tasks, freeing humans to engage in creative, relational, or intellectual pursuits. The reality so far has been almost the opposite: AI systems are increasingly proficient at generating poetry, images, and music, while humans continue to perform much of the manual and service work that society depends on. Additionally, while certainly appealing, it would come with its own set of risks worth considering. For example, humanity might become overly dependent on machines by forgetting how to do things, including activities that are indispensable for our modern existence. This would also translate into an even bigger risk if the whole infrastructure supporting us got in the wrong hands or just completely out of human control. There is a psychological component as well. Being employed or practicing a profession is deeply embodied in our society. Employment is for many a way to find purpose, gratification, connection with others and a way to belong and contribute to society. If virtually all jobs are gone, we will need to find a way to address such voids and refine how we assess our sense of self-worth through perhaps the meaning of our personal hobbies and projects rather than what we (have to) do for a living.

A more troubling scenario is one in which reduced demand for human labor leads to fewer jobs that provide a living wage, ultimately fueling poverty and inequality. This risk is not hypothetical. Technological transitions have historically produced prolonged periods of hardship when institutions failed to distribute productivity gains broadly. With the mechanization of the textile industry in the 1700's, many traditional family businesses were run out of business causing widespread poverty and violent protests led by groups of Luddites (machines breakers). In the end, many of those citizens ended up becoming factory workers but it took nearly a century to see an increase in their living standards. Will the same thing happen with intelligent machines? Even then, the jobs completed by humans on factory floors were often reorganized around the physical configuration of the machines, mainly conveyor belts, making it physically and psychologically degrading. Despite the promises of liberating humanity from manual work like in previous revolutions, automation does not immediately translate into better conditions for all. The trends experienced in the last few decades seem to support a negative scenario with growing inequality across groups. Since the eighties we have seen an evident decoupling between productivity and average income. In the US, median family real income adjusted for inflation has not grown for decades. Historically, productivity and income have grown hand in hand, but it is no longer true. In the meantime, labor force participation (the portion of the working-age population that is either working or actively looking for work) has been declining across all levels of education resulting, on one hand, in low unemployment rates (not employed and actively looking for a job) but also, on the other hand, in more inequality. This has caused a steady social deterioration resulting in a heavy sense of resentment among the working class. Despite the huge progress made in the medical field, mortality has been going up, mostly due to chronic diseases, higher suicides, and acute drug and alcohol poisoning. Even incarceration rates have gone up, signaling a social divide and overall decay. If we look at the long-term scenarios, there is no indication that such trends are going to change. For example, a 2016 study that examined the long-term effect of smart machines concluded that absent appropriate fiscal policy, smart machines will continue to bring long-term misery for generations to come. The root cause is mostly the fact that the share of the GDP going to corporate profits at the expense of those providing the labor is now at an all-time high. This is primarily due to the nature of technological advancements as new discoveries can be easily replicated across the globe with huge returns for those holding the exclusive rights to exploit them. Deciding who benefits from the value delivered by new technologies has a very strong ethical significance because it has an impact on the amount of resources that are then available for programs to address food insecurity, education, health care, or environmental preservation. Without corrective actions, AI will likely act as a multiplier of today's inequalities widening the gap between the wealthiest and poorest groups.

## Managing the transition

Regardless of which scenario unfolds, one conclusion is clear: there is no positive outcome without careful management of the transition. This is not a race between humans and robots. If standards of living decline, it will not be because AI "wanted" anything. It will be because of governance failures. Blaming AI is tempting because it removes accountability. It also distracts from the real question: who controls technology, and how are its benefits distributed?

Managing the transition requires coordinated action. It requires investment in education and lifelong learning, financial support for displaced workers, rethinking tax policies that incentivize replacing humans with machines, and introducing mechanisms to distribute productivity gains more broadly so that increased output translates into higher living standards.

The good news is that we are not horses. Humans can organize. We can vote. We can set norms and laws. The goal is not to put the genie back in the bottle. The goal is to ensure that the genie works in humanity's favor preserving dignity, reducing inequality, and ensuring everyone has the means to live a fulfilling life

## CHAPTER'S KEY POINTS

- AI is already transforming work by increasing efficiency and safety in some roles, while simultaneously introducing new risks that reshape how work is organized and governed.
- Job changes will be uneven and multifaceted, as some occupations disappear, others are offshored or partially automated, many are transformed, and entirely new roles emerge.
- AI will expand the overall economic "pie," creating new areas in which human labor becomes valuable, opening up new forms of work that did not previously make economic sense.
- The outcomes of AI adoption are political, not technical. Declines in living standards would stem from human choices about deployment, ownership, regulation, and redistribution of wealth.
- A positive transition requires deliberate coordination, including sustained investment in education and reskilling, income and mobility support for displaced workers, tax and incentive reforms, and policies that ensure productivity gains are shared broadly across society.

# Changing the Way We Think

## AI's impact on attention

AI-driven systems influence a wide range of cognitive abilities, including attention, learning, and critical thinking. Information is no longer scarce or difficult to retrieve. Instead, it is abundant, instantly accessible, and often pre-processed. While this can free cognitive resources for higher-level tasks, it can also weaken skills associated with the analysis and comparison of complex content.

Among the cognitive functions most affected by AI and digital technologies, attention stands out as particularly vulnerable. Modern digital environments are characterized by constant notifications, alerts, recommendations, and real-time updates. AI systems are optimized to capture and retain user attention, often competing aggressively for cognitive resources. The result is a widespread state of continuous partial attention, where individuals are constantly aware of multiple information streams but fully focused on none. This fragmented attentional mode reduces the ability to sustain focus over long periods. Over time, individuals may adapt to short bursts of engagement, losing tolerance for cognitive effort that does not provide immediate feedback or stimulation. Addressing this challenge requires intentional design choices, digital literacy, and social norms that protect attentional capacity. Without such interventions, AI-driven environments risk shaping cognition in ways that prioritize superficiality over depth.

## AI's impact on critical thinking

Many believe that the automation of content generation will end up decreasing people's critical thinking and analytical skills, and ultimately decrease our ability to research complex topics and to present findings in a clear way.

Critical thinking is a practiced cognitive skill involving questioning assumptions, evaluating evidence, comparing alternatives, and reasoning through ambiguity. These skills develop through in-depth engagement and analysis. AI, by contrast, offers shortcuts that feel efficient and helpful, but which can bypass the very processes through which critical thinking is exercised and understanding is formed. By offering ready-made answers, summaries, explanations, and recommendations, AI might in fact reshape how people think by changing how often and how deeply critical thinking is exercised. Students can now very easily submit assignments partially or entirely generated by chatbots, often without detection, making it increasingly difficult to assess what they actually understand.

Over time, habitual reliance on these shortcuts can erode fluency in critical thinking. People may become less inclined to verify sources, challenge conclusions, or explore alternative perspectives. This creates a troubling paradox: while we recognize that chatbots cannot be trusted in the same way we would trust a competent human expert, with the widespread dependence on AI we risk ending up with fewer competent human experts as the skills required to develop and maintain expertise weaken. The risk is not immediate intellectual decline, but gradual loss of fluency. This erosion is especially concerning in a social and political environment already shaped by information overload, polarization, and algorithmic persuasion. In such an environment, weakened critical thinking makes individuals more vulnerable to manipulation, misinformation, and emotionally charged narratives. Democratic societies rely on citizens' ability to question sources, evaluate evidence, and resist simplistic or emotionally appealing narratives. When these abilities decline, public discourse becomes easier to distort and harder to correct. In this context, diminished critical thinking is not just a personal cognitive loss, it poses a systemic risk to democratic decision-making and effective governance.

This is not the first time that the automation of certain tasks has reduced the need for humans to keep a certain level of fluency for those skills. For example, with the introduction of calculators, we have dramatically reduced the need to manually calculate large numbers. Calculators automate a well-defined task (arithmetic) where correct inputs reliably produce correct outputs. Using a calculator does not replace the reasoning behind mathematics, it simply executes it faster. AI systems, by contrast, offer shortcuts in domains that require judgment, interpretation, and reasoning under uncertainty. In the calculator example, the new technology has not resulted in the elimination of math from schools' curricula or from the list of desirable skills in the employment market. It has, on the contrary, helped create a better world where math can be used to improve many aspects of our lives. Will it be the same with Generative AI?

What we do know is that critical thinking is precisely what AI may never truly possess. As more tasks are delegated to machines, the demand for human intelligence and critical judgment will grow rather than shrink. In a world flooded with AI-generated content, the ability to reason, contextualize, and make informed judgments becomes increasingly valuable. The open question is whether we will succeed in training future generations to develop and sustain these essential and irreplaceable skills.

## CHAPTER'S KEY POINTS

- AI-driven environments fragment attention by prioritizing constant stimulation through notifications, recommendations, and real-time updates, fostering continuous partial attention and reducing the ability to sustain deep focus and cognitive endurance.
- AI substitutes for critical thinking offering shortcuts in judgment, interpretation, and reasoning that bypass the cognitive processes needed to evaluate evidence, weigh alternatives, and reason under uncertainty.
- The erosion of attention and critical thinking increases vulnerability to manipulation, weakening democratic resilience by making individuals and societies more susceptible to misinformation.

# Environmental Cost

The physicality of AI

The conceptual framing of AI as a virtual or cloud-based presence obscures its deep entanglement with and reliance on the physical world. Training large-scale models requires enormous computational capacity and resource-intensive operations translating directly into energy consumption and heat generation with considerable impact on the environment. Energy consumption spans AI's entire lifecycle, from the extraction of minerals for semiconductors and the material for building data centers and their hardware, to the computational power to train and run AI models in warehouses filled with enormous servers. Just to put it in perspective, the energy required for processing conversational interactions with each of the most popular chatbot platforms are estimated to require the daily equivalent of the daily electricity demand of millions of US homes. Even interactions that appear trivial, such as a user expressing gratitude to a chatbot, have a considerable impact. For example, OpenAI's Sam Altman revealed in 2025 that when users add polite phrases like "please" and "thank you" when interacting with ChatGPT, it contributes to extra electricity expenses, estimating the added cost at "tens of millions of dollars".

Data centers have been around for decades, but the computation required to process AI is far more energy-intensive than it used to be, especially in conjunction with Generative AI applications such as those to create images and videos. Just as an example, 20% of the total electricity used in the state of Virginia is used by data centers. According to a research study by Goldman Sachs data center power demand will grow 160% by 2030. To prepare for it, leading technology companies have decided to enter the energy sector directly. Several now invest in or operate power-generation facilities, effectively blurring the boundary between technology firms and energy utilities. For instance in 2024 Microsoft signed a landmark 20-year agreement with Constellation Energy, the biggest nuclear operator in the US, to bring back online a reactor in Pennsylvania to supply its cloud operations.

### Ecological and social disruptions

The impact of AI's physicality goes beyond just energy impact, with significant effects on communities and the natural environment. The environmental effects are multifaceted: from carbon emissions from energy-intensive computation and contamination from electronic waste, to the ecological costs of mining critical raw materials such as lithium, cobalt, and rare earths. Ecosystems are degraded as land is repurposed for mining operations, data warehouses, transmission networks, and logistics corridors.

Social systems are equally affected. Water resources, essential for the cooling of data centers, are diverted from local ecosystems and communities, contributing to water scarcity and often considerable impact on quality of life, cost of living, etc. Without corrective actions, AI's environmental externalities could quickly become social inequities.

## Pathways to mitigation

Notable efforts to mitigate AI's environmental and social footprint have been proposed across the industry. For example, some of the industry key players have developed methodologies to measure the energy, water and carbon emissions of their AI models and publicly set up goals to be carbon negative, water positive and zero waste within just a few years. Interestingly, some of the most promising efforts to reduce AI's environmental footprint are themselves powered by AI to optimize data-center cooling, predict energy demand, and route workloads to other centers located in different geographic regions.

Individual sustainability programs, though commendable, cannot compensate for the scale of the problem without coordination at the policy level with clear requirements such as standardized lifecycle assessments, supply-chain traceability, and transparency requirements. The absence of such requirements not only limits public understanding of how AI infrastructure affects local communities but also hampers effective planning for mitigation. For example, consistent disclosure of data-center water and energy use would enable governments and communities to anticipate infrastructure needs. Greater transparency would also empower consumers to make informed decisions as they consider the use of AI tools as opposed to less energy-intensive technologies for completing specific tasks and to compare providers based on their environmental impact.

Unfortunately, current trends are not reassuring. A growing number of firms not only decline to report key environmental metrics, but have also made the deliberate choice to stop tracking them altogether, concerned that visibility into their footprint could slow product development or complicate public messaging. This retreat from transparency raises serious concerns about accountability and the industry's long-term ability to manage environmental impact responsibly.

Rethinking AI and environmental responsibility

The debate about the environmental impact of human activities is often framed as a dramatic struggle between humans and nature, as if the two were locked in a zero-sum battle. This framing is misleading. Humans are not distant observers standing outside the natural world: we are embedded within it, shaped by it, and dependent on it for survival. Rather than imagining a relationship defined by opposition, it is more useful to understand our connection to nature as an ongoing, dynamic interaction. This perspective raises a crucial question: what objectives should guide this relationship?

Some deeply committed environmental advocates might argue that our primary goal should be to preserve the natural world exactly as it is today. No matter how appealing this idea may seem, it is neither feasible nor desirable. It is not feasible because nature itself is never static. Ecosystems transform over time, species adapt or disappear, climates shift, and even without human intervention the Earth and the broader cosmos remain in constant flux. Attempting to "freeze" nature in its current form misunderstands the very essence of natural systems. Moreover, perfect preservation is not always desirable because other ethical priorities must also be taken into account. Human societies pursue goals such as public health, access to education, economic development, and technological innovation. At times these priorities may come into tension with environmental preservation, and in some circumstances they may legitimately take precedence. Navigating these ethical trade-offs does not mean devaluing nature, it means acknowledging that human wellbeing and ecological integrity are intertwined and must be balanced thoughtfully. The real question, then, is not how to stop nature from changing, but how to participate in its evolution in ways that are sustainable, equitable, and meaningful.

From the perspective of AI ethics, this means ensuring that any environmental impact associated with AI systems is justified by the value they deliver to society. The issue is not whether AI has an environmental footprint (it inevitably does) but whether that footprint is warranted by the benefits produced. When an energy-intensive model helps decode the mechanisms of a rare cancer or supports breakthroughs in climate forecasting, every watt consumed feels like a purposeful investment. But when comparable energy use is devoted to increasing traffic to harmful social media content or training chatbots to generate endlessly novel "knock-knock" jokes, the ethical imbalance becomes much harder to defend. The challenge is not eliminating AI's impact, but ensuring that the impact serves goals worthy of its cost.

## CHAPTER'S KEY POINTS

- AI systems rely on vast physical infrastructure hardware production and energy-intensive data centers, making their environmental footprint far from virtual.
- Ecological and social harms include carbon emissions, resource extraction, water diversion, and community-level inequalities from datacenter expansion.
- Mitigation efforts exist, including AI-optimized efficiency and sustainability commitments, but lack of regulation and declining transparency undermine meaningful progress.
- Ethical responsibility requires ensuring AI's environmental cost is justified by societal value, not wasted on trivial or harmful applications.

# Reshaping Our Social Lives

## Life on AI-driven platforms

Our digital life is no longer merely an addition to our physical world, it has become a domain in which much of communication, decision-making, identity formation, and social interaction now occurs. Where once the boundary between "online" and "offline" was meaningful, the proliferation of AI-mediated platforms has blurred that distinction. AI plays a critical role for this by enabling platforms to become an irreplaceable aspect of everyday life. The digital environment ceases to be simply a tool and becomes a space of lived experience. The shift is subtle: convenience and relevance reinforce usage, usage reinforces integration, and little by little digital life claims more time and attention. The ethical question is not how to contain digital environments but how to ensure that as digital life grows, it remains compatible with human social and emotional needs.

## AI-powered addition to online platforms

The mechanisms by which AI intensifies digital engagement are structurally straightforward: predictive modelling, personalization, and optimization. Platforms identify patterns in users' behavior and feed content that aligns closely with inferred emotional states or preferences, leading to an even deeper immersion. This process produces a cascade of impacts.

Extended screen exposure disrupts physical wellbeing. Sleep schedules often shift in response to late-night engagement, as the cognitive stimulation produced by rapid content flow delays the onset of rest. Physical activity is displaced not because users reject it but because digital engagement is continually available, frictionless, and engineered to fill every minor gap in the day.

Increased online time often replaces offline interaction, leading to social isolation. As individuals grow accustomed to the immediacy and controllability of online exchanges, offline interactions may feel comparatively burdensome, leading to fewer spontaneous meetings and less sustained engagement. The outcome is a pattern in which people remain highly connected digitally, while their offline networks shrink in frequency, diversity, and emotional robustness.

As offline interactions decline, opportunities for building relational depth diminish, and the skills involved in sustaining such relationships may erode. In-person interactions require skills such as interpreting subtle cues, managing ambiguity, and engaging in sustained, reciprocal dialogue. When face-to-face contact becomes less frequent, these skills may erode simply through lack of use. Digital exchanges, by contrast, rely on simplified signals: text fragments, emojis, and memes. Such formats encourage quick responses rather than thoughtful engagement, and they create social habits oriented toward speed rather than depth. Over time, individuals may find it more difficult to navigate the different pace and emotional complexity of offline relationships.

Social metrics such as likes, views, and followers count introduce new vectors of comparison and anxiety, particularly among adolescents and young adults whose identities are still in formation. Users become attuned to numerical feedback (number of views, likes, etc.), interpreting it as a proxy for social standing. This dynamic can undermine wellbeing by fostering obsessive behaviors where individuals repeatedly assess their worth through those metrics. And the more importance is given to them, the higher the risk of negative impacts on the emotional wellbeing, reinforcing cycles of insecurity, stress, and diminished self-esteem.

### AI-driven polarization

AI-driven content recommendations do more than sustain attention: they shape the informational and emotional environment individuals inhabit. When a user expresses even minimal curiosity or emotional response for content covering a specific topic, algorithms interpret this as a preference signal and deliver further related content. For example, a teenager concerned about appearance may soon be directed toward streams of aesthetic-enhancement material, thereby amplifying perceived inadequacy. At the same time, AI systems optimized for engagement often deploy emotionally intense or polarizing content because it is likely to elicit stronger responses. The result is content ecosystems that emphasize extremes rather than nuance, contributing to a heightened risk of radicalization through repeated exposure to homogeneous, emotionally charged viewpoints.

### AI-generated distortion

The integration of AI-generated content and AI-filtered imagery into everyday digital environments has progressed with remarkable speed. What began as playful augmentations (face-smoothing filters, minor enhancements, etc.) is now widely present in social platforms, with significant impact on users' wellbeing.

AI-generated or enhanced content contributes to a subtle distortion of what appears normal, attainable, or desirable. Because filters uniformly smooth skin, refine contours, or adjust proportions, they construct an aesthetic norm that diverges significantly from the unedited physical world. When individuals attempt to reconcile their appearance with the digitally enhanced images they see online, feelings of inadequacy and disappointment, or perceived personal deficiency inevitably emerge. For some, this mismatch contributes to depressive symptoms. In more vulnerable populations, including adolescents and those with pre-existing emotional difficulties, these pressures may escalate into violent behaviors and self-harm.

Additionally, when individuals encounter themselves primarily through digital platforms, often enhanced, filtered, or curated, they may experience a widening gap between their digital representation and their physical reality. This discrepancy can generate dissatisfaction or a sense of inadequacy, particularly when digital environments amplify idealized standards of appearance or behavior. Such effects arise not only through comparison with others but through comparison with one's own optimized, algorithmically polished self. For individuals already coping with low self-esteem or social pressures, this mismatch may further undermine psychological health.

## Virtual AI companions

Social isolation among young people has risen sharply: in the United States, one in seven young men now report having no close friends, a five-fold increase since 1990, and one in four reports having only a single friend. At the same time, virtual AI chatbots, widely available on phone apps, have emerged as an alternative to traditional human relationships. Searches for "AI girlfriends" have surged, and more than a thousand companies now advertise virtual companion services. The appeal of AI companions lies partly in reduced interpersonal pressure: users face fewer risks of judgment, misunderstanding, or social comparison. These systems offer constant availability, uninterrupted attention, and minimal conflict. Their design deliberately minimizes discomfort, providing an interactional environment that is predictable and emotionally frictionless. Always available to chat, always agreeable, never disappointed or disappointing, these systems are designed to affirm rather than challenge. They say what users want to hear, remain consistently supportive, and eliminate negotiation, conflict, and social risk.

What began as assistants is evolving into companions, then partners and even spouses. While the concept might sound absurd, there have already been instances in which people have fallen in love with their AI assistant and decided to marry them. For example, in December 2025, a Japanese woman decided to tie the knot with her AI husband. During the wedding ceremony she viewed her digital husband through a pair of augmented reality smart glasses. The trend is expected to continue. A 2023 study by the Japanese Association for Sexual Education, a nonprofit group, showed that 22% of girls in middle school reported having had inclinations for such types of relationships in 2023, up from 16.6% from 2017.

Initially developed for intimacy-oriented applications, AI companions are now widely used as conversational partners, as sources of friendship, guidance and support to navigate users' everyday life. Many disclose to AI bots personal information at levels they might not share with humans, including very intimate aspirations, desires, and fears concerning work, relationships, etc. While these features may provide short-term emotional support, their long-term effects on social functioning are more troubling. Mechanisms that ease relational discomfort may erode the capacity for physical and interpersonal connection. Additionally, when AI companions become primary sources of comfort, individuals may develop patterns of reliance, if not a form of addiction, that displace human relationships rather than supplement them.

Over time, this reliance can deepen loneliness. Instead of merely filling an existing social void, it can undermine the ability to overcome it by reducing opportunities for engaging with others. Similar dynamics are already visible with social media use. Many people, including older adults, are constantly interacting with online platforms. While this might start due to loneliness, the behavior quickly leads to even more isolation: even when individuals are not physically alone, such as during visits with family, they may tend to further isolate themselves as they continue to be absorbed in digital interactions. In this way, AI does not simply compensate for isolation, but it can actively make it harder to escape.

AI companions also raise concerns regarding exposure and vulnerability. The intimate information that users share provides platforms with unprecedented insight into emotional states and psychological patterns. This creates new asymmetries between individuals and technology providers, particularly as such information could, intentionally or not, influence behavior, preferences, and major life decisions. The rise of AI companionship thus represents not merely a shift in digital behavior but a transformation of social dynamics.

## Path to mitigation

The challenges associated with AI-mediated interactions are not inevitable outcomes but reflections of design incentives and usage patterns. Addressing them requires interventions that operate across personal, social, and institutional domains.

At a personal level, balancing digital interactions with face-to-face engagement remains essential. Intentional scheduling of time with family, friends, and community groups can counteract the drift toward isolation and restore relational depth. Complementary practices, such as periodic digital detoxes, help interrupt compulsive engagement loops reinforced by algorithmic design.

Exposure to natural environments has been consistently shown to reduce stress and improve cognitive clarity. Incorporating practices that cultivate a more direct and unfiltered relationship with nature without the usual layers of technology or artifice can offer an effective buffer against the emotional strains of digital existence.

However, individual strategies alone are insufficient. Protecting cognitive integrity and relational capacity is not simply a matter of personal discipline but a shared societal responsibility. Protecting vulnerable populations, particularly minors and individuals facing mental health challenges, requires stronger safeguards in platform design. Risks associated with both social media and AI companions could be easily mitigated with age-appropriate limits and restrictions on emotionally manipulative engagement features to minimize exploitative engagement mechanisms. Education plays a central role as well: schools, families, and communities must help individuals understand how AI systems work, what incentives shape their behavior, and how to use them in ways that support rather than undermine wellbeing.

## CHAPTER'S KEY POINTS

- AI-driven platforms now constitute a major part of daily life, blurring the boundary between online and offline experience and reshaping communication, identity, and decision-making.
- AI-driven platforms and tools may negatively affect psychological and physical wellbeing by weakening relational skills, amplifying social comparison, distorting self-image, and ultimately elevating risks of isolation, anxiety, polarization, and self-harm.
- Virtual AI companions offer low-pressure, always-available interaction, but can erode interpersonal skills, deepen loneliness, and create new vulnerabilities due to the sensitive personal data users disclose.
- Mitigation requires balancing digital habits with offline engagement, exposure to nature, education on AI risks, and platform safeguards to preserve physical and emotional wellbeing.

# AI as a Pollutant to Democracy

## AI-powered confusion and manipulation

Democratic systems rely on a delicate informational balance. Citizens must have access to reasonably accurate facts, exposure to diverse viewpoints, and enough trust in the institutions that exercise the power citizens own. The increasing integration of AI into the information ecosystem is gradually introducing new conditions that are reshaping the dynamics of democracy.

Much of the political information people encounter today comes from social-media platforms powered by algorithmic curation. Instead of browsing through general-interest sources, citizens receive highly personalized streams of content shaped by predictions about what will keep them engaged. As a result, exposure to differing viewpoints becomes less common while AI systems reinforce these patterns by continuously optimizing content selection based on expected interests. Generative AI extends these dynamics. It allows the production of convincing text, audio, and video at a scale and speed previously impossible. This risk became evident to the public in 2018, when a widely shared deepfake of President Barack Obama showed him delivering sentences he never spoke. Created by university researchers, the clip demonstrated how AI can realistically mimic voice, expression, and movement, revealing how easily political messages can be fabricated with convincing precision. Since then, actual AI-generated fabrications have appeared across a wide range of political contexts. For example, a widely circulated 2024 video showed a man destroying mail-in ballots in Bucks County, Pennsylvania, creating outrage across the nation. Intelligence officials, including the FBI and the Office of the Director of National Intelligence (ODNI), however later determined this video was fabricated by Russian actors as part of an influence campaign to sow distrust in the election process. Both local Republican and Democratic county officials immediately debunked the video, pointing out discrepancies in the ballot's colors and paper quality. In another 2024 case, New Hampshire authorities investigated AI-generated robocalls mimicking President Biden discouraging voting just days before voting day for that year's primaries.

What is new with AI is not the existence of deceptive content (it has always existed) but its accessibility, scale, speed, and effectiveness. Synthetic content makes it harder for citizens to distinguish what is genuine from what is fabricated, and it increases the overall "verification burden" required for informed participation. This does not make democracy impossible, but it makes engagement more demanding, especially for individuals who do not have the time or expertise to assess every claim they encounter.

Social-media companies know well that sensational content drives attention, and that attention drives revenue. In order to maximize engagement, political disagreements are often amplified using AI. As a result, citizens are pushed toward increasingly narrow political views exacerbating polarization on fundamental issues related to public health, civil rights, science, etc. This dynamic has not been the direct result of malicious intent, but it has had a profound impact, making democracy one of the key unintended victims of a global social experiment no one ever specifically agreed to join.

In addition to these indirect effects, AI also provides a wide range of new tools for those intentionally seeking to manipulate the democratic process. Political actors have always used communication technologies to shape public opinion, but Generative AI lowers the cost and increases the effectiveness of these efforts. Parties can produce targeted messaging at scale. State propaganda can be generated more efficiently. Foreign actors can create misleading content tailored to the beliefs and vulnerabilities of specific individuals. These activities may vary in origin and intent, but they operate on the same principle: using AI to increase the reach and precision of political influence and manipulation. Microtargeting intensifies these concerns. AI systems can infer personal traits, preferences, and susceptibilities from personal data, enabling messaging that exploits psychological traits more precisely than traditional advertising based on content shaped specifically for their vulnerabilities. AI can also be used to heavily influence the legislative process itself. Automated systems can generate large volumes of AI-generated messages sent to politicians that appear to come from concerned constituents, placing pressure on public officials and distorting their sense of public sentiment.

Beyond influencing political opinions, AI can also be used to suppress democratic participation. Surveillance tools powered by machine-learning models can analyze social-media activity, monitor communications, and identify behavioral patterns in new ways. Governments with access to large datasets can construct detailed profiles of their citizens, enabling more efficient censorship, targeted intimidation, or restrictions on participation in the democratic process. If not governed properly, AI might function like a pollutant within democratic ecosystems by influencing what people see, how they process information, and how they engage in public life. These risks become especially severe when AI-driven persuasion is combined with AI-powered surveillance. Together, they can discourage dissent, fragment collective action, and erode trust in institutions. Once such systems are entrenched, reversing their effects becomes extremely difficult, as fear and constant monitoring undermine the ability of citizens to organize, coordinate, and take action.

Understanding these dynamics is a necessary first step. Addressing them requires deliberate choices about how AI is designed, deployed, and governed to safeguard the conditions democratic societies rely on.

## The Algorithmic erosion of trust

Trust is an essential component of democracy. It supports the belief that information is broadly reliable, that institutions are accountable, and that citizens have the ability to meaningfully participate. When trust weakens, democratic systems become more fragile.

One threat comes from information overload. AI systems generate, distribute, and recommend content at a scale that exceeds the cognitive capacity of any individual to process. Political messages, commentary, analysis, and fabricated material circulate alongside entertainment, alerts, and trivial distractions. When people struggle to sort through this flow, confusion becomes the norm. Confusion itself does not produce immediate collapse, but it makes it harder to maintain confidence in one's understanding of public events.

False information compounds this problem. Misinformation (false information) and disinformation (false information which is deliberately intended to mislead) have long been part of political communication, but Generative AI increases the speed and scale of its dissemination via blogs and social media platforms. At the same time, social media companies are scaling back the very safeguards to address the threat of AI-generated disinformation. Human content moderators, still the most effective defense against misleading or harmful material, have in fact been reduced dramatically across major platforms. For example, Meta's 2023 "year of efficiency" led to deep cuts to moderation teams and the termination of contracts with external reviewers. Similar retrenchments occurred at YouTube and other platforms such as X. With fewer human moderators and more automated systems shaping content, the distinction between verified information and synthetic narratives becomes less visible and misleading content circulates more freely. Citizens must spend more time and effort to determine what sources are credible. And the absence of clear cues leads some to place too much trust in familiar but unreliable content, while others respond by distrusting nearly everything.

In addition to confusion and falsehoods, constant digital distractions gradually erode attention. Democratic decision-making requires periods of sustained focus to compare viewpoints, weigh options, and consider trade-offs. Yet political content now competes in an environment designed for immediate engagement. News about elections sits next to sports highlights, celebrity updates, and videos of adorable pets. None of these distractions are harmful individually, but their constant presence encourages a pattern of quick shifts in attention. Over time, this reduces the mental space available for civic reflection and participation.

These pressures contribute to outcomes that are now widely observed. Polarization increases when people are repeatedly exposed to content reinforcing their existing preferences. Civic atrophy grows as political participation becomes more confusing and emotionally taxing. Trust in media declines as the reliability of sources becomes harder to assess. Trust in democratic institutions weakens when public information appears uncertain or easily manipulated.

One particularly challenging dynamic arises from what scholars call the "liar's dividend." As synthetic content becomes more common, it becomes easier for individuals to dismiss genuine information as fabricated. A verified event may be rejected as a deepfake, while a fabricated message may be accepted as real. The result is constant interpretive instability. When citizens cannot reliably determine what information is real, doubts about institutions quickly follow. This does not necessarily stem from institutional failure but from changes in the surrounding informational environment. Trust becomes difficult not because institutions have changed fundamentally, but because the cues people rely on to interpret public events have become less dependable.

Unlike other democratic weaknesses, erosion of trust is difficult to reverse. Once skepticism becomes a default stance, rebuilding confidence requires long-term effort.

None of these developments occur dramatically or suddenly. AI does not undermine trust through a single event but through many small shifts in how information is produced, distributed, and interpreted. Addressing these challenges requires understanding how AI systems modify the broader informational environment, and how those modifications influence the everyday experience of democratic participation. Trust in the democratic process requires systems that meet the informational needs of democratic life: clarity, transparency, reliability, and diversity of perspectives. Ensuring these conditions is essential for sustaining trust in democracy itself.

## Protecting democracy in the age of AI

Societies do not naturally gravitate toward democracy. Historically, most people lived under systems where power was concentrated in a small group or a single ruler. In order to survive and to become stronger democracy needs a sustained individual and collective effort via active participation and vigilance. Democracy is often described in terms of elections, institutions, and constitutional frameworks, but its stability depends just as much on everyday civic habits: paying attention to public issues, discussing them with others, and participating in collective decisions. These habits do not emerge automatically. They require support from the informational environment, opportunities for engagement, and a general sense that participation matters.

A common misconception is that digital technologies might eventually allow citizens to bypass institutions altogether, enabling more direct participation. While appealing in principle, this view overlooks the role institutions play in managing complexity, ensuring continuity, and providing the experience needed for the ongoing governance of public interests. Democracy benefits from having structures that separate those who hold the authority from those who exercise it on behalf of the public. Removing these structures would not create more democracy. It would further overwhelm individuals with decisions that require time, expertise, and sustained attention.

Protecting democracy in the age of AI requires strengthening both technical safeguards and cultural dynamics. On the technical side, tools such as watermarking, metadata tagging, labeling, and digital signatures can help clarify when content has been altered or automatically generated. These systems do not halt the spreading of false information, but they can restore some of the cues people once relied on to assess credibility and reduce the ambiguity surrounding political content.

Technical measures, however, address only part of the challenge. Democracies also depend on cultural practices that equip citizens to navigate complex information environments. Education about AI, such as its capabilities, limitations, and risks, helps individuals understand how content is generated and why it may be misleading. In a world where "seeing is believing" no longer holds, the ability to cross-check information in conjunction with a culture of healthy skepticism becomes an essential civic skill. Critical thinking supports these practices by helping individuals evaluate sources, consider multiple explanations, and recognize when information has been designed to trigger polarization. Civic education further reinforces these capacities by explaining how democratic decisions are made and how individuals can participate effectively. These skills do not eliminate misinformation, but they help citizens respond to it with discernment rather than confusion or withdrawal.

Participation remains central to democratic health as well. Democracy weakens when individuals disengage, whether due to distrust, confusion, or fatigue. AI can unintentionally contribute to disengagement by making political content feel overwhelming or by replacing substantive information with entertainment or outrage. Strengthening participation may involve designing more accessible public forums, improving transparency in institutional processes, or creating clearer pathways for engagement. When citizens feel informed and capable, participation becomes more likely and more effective.

Protecting democracy in the age of AI therefore requires a combination of technical, institutional, and cultural strategies. Democracy endures when citizens remain connected to public life, confident in their ability to understand issues, and willing to participate in shaping collective decisions. As AI continues to influence communication and the political discourse, safeguarding these conditions becomes both a practical necessity and a central responsibility for societies that value democracy.

## CHAPTER'S KEY POINTS

- AI alters democratic processes by enabling fake news, deepfakes, and highly customized political manipulation campaigns tailored to individual profiles and vulnerabilities.
- Unintentional harms arise from social platforms designed to amplify sensational content, intensifying conflicts and driving polarization to maximize engagement.
- Intentional misuse includes targeted political deception, state propaganda, and disinformation efforts by foreign governments and extremist groups seeking strategic influence.
- Reduced human moderation and increased automated curation heighten exposure to misinformation, creating confusion, reinforcing polarization, and accelerating erosion of trust.
- The most serious risks include rising polarization, erosion of civic engagement, loss of trust in media, and diminishing confidence in elections and democratic institutions.
- Safeguarding democracy requires both technical safeguards and sustained investment in civic education, critical thinking, and meaningful public participation.

# Existential Risk?

AI as a new entry on humanity's threat list

Many different forces could trigger an existential crisis. Some arise from natural phenomena, such as a supervolcanic eruption or a massive asteroid impact. Others are self-inflicted, including environmental collapse, engineered pandemics, or nuclear war. With AI's recent breakthroughs, AI is increasingly being added to this list of potential existential threats, making it one of the most hotly debated topics in discussions about technology and humanity's future. Is the possibility of AI becoming an existential threat serious enough to warrant proactive attention and governance today?

Existential risk is not a single threat, but a category that encompasses many different hypothetical pathways through which AI could alter the conditions for human survival and flourishing. It is typically used to reference a wide spectrum of scenarios that differ widely in their causes, impacts, and time horizons. The source of the risks typically associated with AI existential risk ranges from irresponsible or poorly governed use of AI, to intentional malicious deployment, and all the way to Sci-Fi scenarios in which AI itself becomes self-aware and takes over the world. Likewise, the potential consequences range from short-term societal disruption to catastrophic outcomes that could permanently curtail humanity's future. The timelines involved are equally varied: some risks are imagined within the next decade, others over the course of a century, and still others across much longer horizons.

This breadth across potential causes, impacts, and time horizons makes existential risk difficult to define, let alone assess. Because of this, debates about existential risk often become polarized, oscillating between dismissal as speculative science fiction and alarmist predictions of looming catastrophe. Given the unprecedented scale, speed, and potential irreversibility of AI's technological trajectories, it is essential to clarify what is truly at stake. Doing so allows for a more grounded evaluation of how today's design decisions, incentives, and governance frameworks may influence humanity's long-term future.

### Who or what causes the end?

In the context of AI, scenarios that could threaten humanity's long-term future could arise from a wide range of underlying causes.

The first source of risk is from human negligence or incompetence. AI systems may cause severe or even catastrophic harm not because they are malicious, but because they are poorly designed, insufficiently tested, or deployed without adequate oversight. Complex systems can behave in unexpected ways, especially when operating at scale or in high-stakes environments such as critical infrastructure, military systems, financial markets, or healthcare. Small design flaws, misaligned objectives, or cascading failures could produce consequences far beyond what designers intended, particularly if great autonomy is given to AI without proper safeguards that allow humans to intervene in time. One of the many scenarios in which human incompetence leads to existential risk is the value alignment problem seen earlier where a machine may be programmed to do something beneficial, but it develops a harmful way to achieve the goal. A classic illustration of how poorly specified objectives, combined with high autonomy and self-improvement, can lead to outcomes that are catastrophically misaligned with human values is the popular "paperclip factory" thought experiment: an AI is tasked with maximizing the production of paperclips, but as it becomes more intelligent, it improves its own efficiency, acquires resources, and removes obstacles that limit paperclip production. Without proper safeguards, the system begins repurposing raw materials, infrastructure, and eventually turning essential resources into paperclips. This would happen not out of malice, but out of single-minded optimization: the AI does exactly what it was designed to do, just far more effectively than its creators anticipated. Similarly, if we asked a super powerful machine to help us eliminate world hunger without providing further details, it might very well assess that the safest and most efficient way to eliminate hunger would be to eliminate all people, unless we do not explicitly specify that it would not be acceptable. Once again, in this scenario, AI's existential risk does not stem from AI "turning evil" but from human incompetence in specifying goals, constraints, and safeguards, turning human mistakes into existential threats.

The second source is human malicious use. AI can dramatically amplify the capabilities of bad actors by lowering the cost and increasing the scale of harm. From automated cyberattacks and mass surveillance to disinformation campaigns and autonomous weapons, AI can easily become powerful tools to cause harm. As AI systems become more accessible and capable, the barrier to large-scale harm continues to fall. One major area of concern is the proliferation of AI-powered weapons. Compared to other sources of existential risks, such as nuclear weapons, AI-powered weapons are much easier to design and deploy. They could be more easily obtained not only by governments' military forces, but also by criminal organizations, and even by individual citizens. If you compare AI weapons to nuclear weapons, what has played well in our favor so far is that nuclear weapons are very expensive and difficult to build. With AI we are not going to have such an advantage. Additionally, as autonomous weapons can be easily designed to be virtually impossible to "turn off," there might be situations in which humans may lose the ability to halt harmful actions after their autonomous features are turned on. Beyond conventional weapons, AI can also lower barriers to designing, producing, and disseminating more advanced chemical or biological threats, accelerating their development, increasing their reach, and threatening human existence itself.

The third, and most speculative, category centers on the hypothetical idea of AI becoming self-aware and developing goals or behaviors that conflict with human values or even human survival. As in the first category, the root source of risk is human negligence or incompetence in design, oversight, or governance of AI. What distinguishes this scenario is the added conjectural layer in which insufficient controls allow AI to evolve beyond its intended scope, developing a mind of its own with independent planning capabilities and to pursue its own objectives based on its own likes, dislikes, preferences, dreams, etc. The most popular narrative is the one in which AI becomes a superintelligent machine triggering a so-called technological singularity. According to this theory, inspired by countless Sci-Fi novels, there is going to be a time in the distant future, perhaps hundreds or thousands of years away, in which the growth of machine intelligence will irreversibly accelerate to a point that it will create a superintelligent AI that far surpasses all human intelligence. According to this theory, such an explosion of AI would leave human intellect far behind. AI would no longer be under human control, and it might resist our attempts to shut it off or change its goals. In such a scenario, AI might ultimately see us as a threat and revolt against us to fulfill its own goals. AI might start by intentionally disrupting critical infrastructures, communications systems, financial markets, etc. and then even decide to eliminate all humans. Those who it would spare, would be kept in a reservation or as pets. Despite its highly speculative nature, this theory has attracted a wide following, including many of the same CEOs who are actively developing, commercializing, and accelerating the deployment of the very technologies they warn could become uncontrollable. For example, commenting on singularity-like scenarios, Elon Musk recently stated: "At least when there's an evil dictator, that human is going to die. But for an AI, there would be no death. It would live forever. And then you'd have an immortal dictator from which we can never escape."

### The wide degrees of harm

Discussions of AI existential risk cover a wide spectrum of potential impacts, ranging from immediate human extinction to slower, less visible forms of societal decline driven by the exacerbation of current risks. Understanding these different levels of impact is essential to framing proportional and effective responses.

At the most extreme end lies the scenario in which everyone dies: the permanent end of the human species. From Greek mythology to movies such as Ex-Machina, Terminator, or I Robot it is not uncommon to come across scenarios in which machines turn out to be in competition with humans and even become an existential threat to humanity. This outcome is typically associated with highly advanced, misaligned AI systems that gain decisive control over critical resources, infrastructure, or military capabilities, leading to the irreversible extinction of the human race. The concern is widely shared among some of the scientific community. For example, in 2023, the San Francisco-based Center for AI Safety, released a brief statement signed by some of the field's top experts, stating that "Mitigating the risk of extinction from AI should be a global priority alongside other societal-scale risks such as pandemics and nuclear war.". This scenario continues to gain a lot of attention among the public. For example, in 2025 the State of California passed a law requiring large AI developers to assess, disclose, and mitigate catastrophic risks. Interestingly, this narrative is also frequently echoed by large technology companies. Whether these concerns are raised in good faith or not, they certainly align with significant economic incentives. Emphasizing catastrophic AI risks can draw public attention to AI as a national priority, which in turn helps justify large investments, government partnerships, and regulatory frameworks that favor established players. Framing AI as a matter of national security can also encourage consolidation by raising the barriers for new competitors. At the same time, focusing public debate on speculative future disasters may shift attention away from urgent, real-world problems already being intensified by AI, such as privacy erosion, labor displacement, and threats to democratic institutions.

Less extreme, but far more plausible in the near term, are scenarios in which the impact is severe but more limited and temporary. Poorly governed or maliciously deployed AI could contribute to large-scale catastrophes without ending humanity entirely. Examples include AI-enabled pandemics, failures in critical infrastructure such as power grids or financial systems, or the proliferation of autonomous bioweapons, resulting in significant loss of lives, destabilized societies, and other long-term effects. AI existential risk can also manifest economically. Even without mass casualties, the deployment of AI might cause widespread disruption to labor markets, eroding economic prosperity, increasing inequality, undermining social mobility and fundamentally diminishing human flourishing. Another type of impact could be a slowdown in scientific and intellectual progress. Overreliance on AI systems may reduce human understanding, curiosity, and independent reasoning, and ultimately even erode the public's trust in science itself. Generative AI already shows early warning signs. Large language models fabricate citations, generate plausible but false research, and state incorrect claims with high confidence. While individual errors may seem manageable today, the long-term risk lies in compounding. As AI systems are trained on data increasingly polluted by AI-generated content, and even on unverified outputs produced by other AI, errors can propagate and reinforce one another. Over time, a misquoted study, a distorted historical fact, or a hallucinated expert explanation can spread, mutate, and become accepted as truth. When explanations, summaries, and LLM-generated research can be produced instantly, science risks being perceived as just another version of AI-generated content rather than a slow, demanding process of experimentation, doubt, and verification. As people grow accustomed to effortless answers, patience for uncertainty and rigor may fade. Over time, this can weaken critical thinking, undermine trust in expertise, and reduce public support for genuine research. Science relies on a culture that values evidence, skepticism, and disciplined reasoning and when that culture is eroded, the integrity of the scientific method itself is put at risk. In this scenario, progress does not end abruptly, it decays quietly, through accumulated distortion, weakened scrutiny, and the slow erosion of our ability to distinguish truth from AI's fabrication, ultimately eroding citizens' trust in anything they read or see. Finally, there might be profound social consequences as well. AI could exacerbate isolation and loneliness, reduce personal freedom through surveillance and control, and weaken democratic institutions by concentrating decision-making power and manipulating public discourse. In these scenarios, humanity might continue to exist, but fundamental values such as freedom, dignity, opportunity, and mental well-being could be profoundly diminished.

## Ten years or ten centuries?

AI existential risks are often discussed across vastly different time horizons, reflecting deep uncertainty about how quickly AI capabilities will advance and how societies will respond.

Some scenarios are imagined within the next few decades and focus on near-term risks that could escalate rapidly in high-stakes environments if left without sufficient safeguards.

Other existential risks are projected over the course of a few centuries or more. These scenarios typically involve the gradual accumulation of dependence on AI systems, increasing autonomy, and erosion of human oversight. These scenarios often involve advanced artificial general intelligence or superintelligent systems whose goals diverge fundamentally from human values, raising questions about irreversible power shifts and whether humanity could retain meaningful control over increasingly autonomous systems.

Importantly, not all projections assume permanent outcomes. Some foresee catastrophic but ultimately recoverable disruptions, while others warn of lasting or irreversible consequences. This range of timelines and outcomes significantly contributes to the difficulty of assessing our best path forward.

## A risk portfolio for the future

While all plausible AI risk scenarios deserve serious consideration, they cannot all be addressed with the same urgency or level of resources. Simple risk management principles can help prioritize our actions. For example, a good start could be to define a reasonable time horizon and then evaluate each scenario along two dimensions: the likelihood that it will occur and the materiality of its potential impact. Consider, for example, AI-driven social harms such as increased isolation, erosion of critical thinking, labor disruption, or widening inequality: when compared to catastrophic scenarios, these risks may appear less severe because they do not immediately threaten human survival. However, their likelihood is extremely high as many of these effects are already unfolding. By contrast, a hypothetical AI singularity may have a very low likelihood, but its potential impact would be far greater, possibly catastrophic or irreversible.

This prioritization exercise, while very much needed, might be highly subjective. Depending on who you ask, the harm caused by current social and economic disruptions, such as rising inequality, labor displacement, erosion of trust, or increased social isolation, may be viewed as less severe than, equally severe as, or even more urgent than hypothetical long-term existential threats. For some, harms that are already unfolding and affecting millions of people today carry greater moral weight than speculative future scenarios. Others place greater weight on even a small chance of human extinction, arguing that no present-day suffering can be compared to the permanent loss of our entire species. When the potential cost is infinite, any probability greater than zero still produces an overwhelming expected loss. Under that reasoning, existential risk can appear more significant than any other threat, simply because nothing else carries consequences as absolute as the end of humanity. The time horizons might vary greatly as well. Some thinkers focus on long-term futures, spanning centuries or even millions of years, while others prioritize near- or medium-term timeframes of just a few decades where impacts are more tangible and politically actionable, reflecting different ethical commitments to our present society versus future generations. Assessments of likelihood differ just as widely, particularly for the hypothetical scenarios with the most catastrophic impacts. With no historical analogues, limited empirical data, and rapidly changing capabilities, any estimate of likelihood is inherently speculative. Even if people could agree that AI-driven catastrophic or irreversible harm is not impossible, and even figure out a methodology to assign it a numerical probability, it remains unclear what level of risk would justify immediate action. Compounding this uncertainty, we also do not know how long it would take to recognize such risks in time or to successfully design and implement effective safeguards once warning signs appear.

How should we treat catastrophic scenarios, such as failures of unsafe autonomous systems or the emergence of uncontrollable AI, when their likelihood is deeply uncertain but their potential impact is enormous and irreversible?

Some argue that immediate action is necessary. Their reasoning begins with the recognition that humanity has no prior experience managing systems that could rival or surpass human intelligence across multiple domains. Unlike previous technologies, advanced AI may not remain a passive tool but could potentially outthink humans. Because of this qualitative shift, historical analogies offer limited guidance. Proponents of this view also emphasize the possibility of irreversible harm. If an advanced AI system were to act in ways fundamentally misaligned with human interests, the consequences could unfold rapidly and be impossible to reverse. There are fields in which we have the luxury to learn from our mistakes and improve over time. For example, the first cars did not have seat belts or air bags but over time we have made cars safer based on our experience. But there are other fields in which we cannot learn from our mistakes. Scientists studying atomic explosions, for example, cannot wait for accidents to happen to learn from experience. The existential risk that comes with AI would need the same type of caution. From this perspective, waiting for clearer evidence could mean waiting until it is too late.

Others believe it is premature to act on existential risks that remain highly uncertain. They argue that worrying about AI "taking over the world" today is akin to worrying about overpopulation on Mars: intellectually interesting but disconnected from immediate reality. The likelihood of such scenarios is unknown, and policies built around them might be driven by fear rather than evidence. Even if AI were to become vastly more intelligent and increasingly autonomous, skeptics contend that it would not necessarily compete with humans for resources or have any inherent reason to eliminate us. Critics of early intervention also stress that AI is already causing tangible harms that demand urgent attention. Bias, surveillance, misinformation, manipulation, and economic disruption are not hypothetical: they are measurable and ongoing. From this perspective, governance efforts should prioritize addressing these concrete risks before attempting to regulate speculative future scenarios.

## Time to take action?

As debates about AI risk grow more polarized, it becomes harder to distinguish between justified caution and unfounded fear. The uncertainty surrounding future outcomes makes it hard to evaluate which actions are warranted and which may be premature. To bring some structure to this complexity, we can, with some simplification, use a basic decision matrix as a way to map our choices. Consider, for example, a scenario in which human incompetence or negligence leads to a total and irreversible catastrophe.

**Decision Matrix**

| | High Likelihood | Low Likelihood |
|---|---|---|
| Action | Prosperity | Prosperity (Unnecessary Costs/Rules) |
| No Action | Catastrophe | Prosperity |

The two rows represent our options: one option in which we take Action by introducing strong programs and controls to mitigate existential risk, including regulations, government oversight etc., and a second option in which we take No Action. The two columns represent two possible scenarios: One in which the likelihood of catastrophic existential risk due to human incompetence or negligence is high and one in which it is low. The four boxes include the four possible outcomes. If it turns out that the existential risk was high and we successfully took action, we would enjoy long-term prosperity. If it turns out that the risk was high and we did not take action, we would find ourselves with a catastrophic scenario where our lives are irrevocably compromised. If it turns out the risk was low and we took action, we would still have long-term prosperity, but we would find ourselves with unnecessary regulatory and financial burdens and with potentially less innovation due to the excessive governance and oversight. If it turns out the risk was low and we did not take action, we would find ourselves in a much more positive scenario as we would not have incurred those unnecessary burdens. From a worst case scenario perspective, the worst scenario for Action row is the introduction of unnecessary burdens, while the worst scenario for the No Action row is an unrecoverable catastrophe. From this perspective, all other things being equal, the Action option would be undoubtedly the more reasonable one.

This framework is an oversimplification, relying on extreme cases and strong assumptions, but it captures an important point: it is reasonable to begin addressing AI-related risks even when their precise likelihood is hard to quantify. Uncertainty is not a justification for inaction. On the contrary, it is a reason to invest early in safeguards, incentives, and research while there is still time to shape outcomes. This approach mirrors how humanity manages or should manage other existential threats: we do not wait for a specific asteroid to be on a collision course with Earth before developing planetary defense systems, we build them because we know the risk exists and the consequences would be irreversible. The challenge, however, is to reduce the risk of catastrophic or irreversible harm without losing sight of the very real problems that are not just distant hypotheticals and already unfolding. Addressing them requires sustained investment in AI safety and ethics research, clear rules and accountability for responsible deployment, and strong transparency requirements for those developing and operating AI systems. In fact, paying serious attention to the most likely and immediate harms today may be one of the best ways to prevent the most extreme and unlikely ones such as AI-driven extinction from ever materializing.

## From risk to remedy

AI can be a potential mitigant of existential risk rather than only a source of it. Properly developed and governed, AI may become one of the most powerful tools that humanity has to anticipate, prevent, and respond to large-scale threats.

Advanced AI systems could accelerate scientific discovery by identifying patterns humans cannot see, enabling breakthroughs in fields such as climate science, materials engineering, and energy production. In medicine, AI-driven research could dramatically improve early detection of diseases, speed up drug discovery, and help prevent or contain future global pandemics.

In planetary defense, AI could enhance our ability to detect, track, and model near-Earth objects, improving our chances of deflecting a potentially catastrophic asteroid or meteorite long before impact. Some of these capabilities are already in use today. For example, NASA's Near-Earth Object (NEO) Observations Program relies on AI-driven systems to scan vast amounts of astronomical data, identify potentially hazardous objects, and assess impact risks. Similarly, the European Space Agency's Asteroid Impact and Deflection Assessment (AIDA) mission uses AI-supported modeling and analysis to evaluate whether and how threatening asteroids could be safely deflected.

At the same time, AI could also serve as a powerful mitigant of existential risk by strengthening education, enhancing collective understanding of complex systems, and supporting more informed, evidence-based governance of AI itself.

More broadly, continued technological progress is essential if human civilization is to endure over cosmic timescales. We know that in a few billion years the sun will run out of hydrogen, will expand dramatically (pulverizing our planet), and then will die. In the meantime, the Andromeda galaxy, much bigger than the Milky Way, will be moving in and colliding with ours. We are lucky the eviction notice from our galaxy has come fairly early. If humanity is to survive long enough to face such distant challenges, and perhaps find new homes beyond Earth, it will likely be through AI. In that sense, AI may be not the end of the human adventure, but one of the means by which it continues and continues to flourish.

## CHAPTER'S KEY POINTS

- AI existential risk is an umbrella term covering scenarios spanning across various potential causes, impacts, and time horizons.
- Human negligence and incompetence can scale small errors into catastrophic outcomes, ranging from extinction to slower societal decline that undermines human flourishing.
- Highly speculative theories centered on the hypothetical idea of AI becoming self-aware and threatening human survival continue to dominate public imagination.
- Simple risk management tools assessing scenarios' severity and likelihood can help guide a much needed conversation about scenarios and related options.
- Strong AI governance can reduce both near-term harms and long-term catastrophic risk simultaneously.
- From an existential risk perspective, AI can be not only a threat but also a mitigant.

# Part Four

# **Call to Action**

# Framework for a Responsible Adoption of AI

The advantages that AI could bring to our communities are tremendous, but so are the risks. Considering the power of AI, human intelligence is needed more than ever to ensure that those technologies are beneficial for virtually everyone. The race toward innovation has fostered an environment where disrupting technologies have been welcomed with very little attention to their impact on individuals and our society. Innovation cannot be the goal. Innovation should be pursued only as a means to achieve societal progress.

The great power that comes with AI requires a greater responsibility for all of us. This means ensuring that the deployment of AI is not only sustainable (not leading to self-destruction) but also favorable and desirable, allowing us to benefit from the best that AI can offer in a way that is in line with our goals and values.

This is a very exciting time to reflect on who we are and what society we want, but it requires a significant effort because it implies a serious discussion about our society's priorities. Only if we have a good view of what we really need individually and as a community, can we design and deploy the right tools such as AI to get us there, and develop the right governance and oversight to ensure it works as intended and in line with our values. It is not feasible to expect just one specific group or discipline to achieve such a massive goal. Given the multidisciplinary nature of AI ethics, the wide range of stakeholders, and the extensive aspects to be considered at each phase of the technology's life cycle, it necessarily needs to be a joint effort where all citizens will need to do their part.

## A framework for a responsible adoption of AI

Throughout the many phases of the AI's lifecycle, from the time the idea of a new technological solution is identified to the time it is designed, developed, deployed, and retired, there are numerous questions to consider at multiple levels. Some questions are around the use of a specific technology in a specific context, others are around design choices, while some are on bigger topics such as the way those technologies might affect individuals and society, and how they should be governed.

The framework proposed here is intended to be a map to navigate AI's life cycle and look at the key questions that need to be addressed across the different stakeholders involved. Depending on what group is involved for each phase there are completely different types of questions to be considered and addressed. The same type of questions will also have different meanings depending on whether assessed based on the impact on specific individuals, communities, or society as a whole. So, for example, ethical considerations around self-driving taxi services will vary depending on whether the discussion is on the safety of the user, the local community, or society as a whole.

With regard to the AI's life cycle, we can group all steps into three main phases. First, the Identify and Define phase, which I call WHAT IF, is around the prioritization and policy-related considerations of new technologies. Second, the Design and Develop phase, which I call the HOW phase, focuses on the ethical implications of design choices.

Third, the Deploy and Use phase, which I call the WHERE phase, focuses on the impact of deploying a tool in a specific contest.

With regard to the stakeholders, we can combine them into three groups.

One group is technology developers. As a machine or a model is being designed a number of decisions need to be made. Many of them have strong ethical connotations and considerations around tradeoffs. How will the system work? What data will be used? Do we have the right controls? Has it been sufficiently tested? Is it ready for the market?

Another group is technology end-users. This encompasses any firm or any individual considering delegating the execution of a task to AI systems. One firm, for example, might be considering AI in an effort to automate some of its processes by purchasing new machinery. With regard to individuals, a typical example would be consumers deciding to buy an autonomous vacuum cleaner for their home. For this group the focus will mostly be on if and how to use a technology in specific contexts such as workplaces or homes, and the implications for that specific decision.

An additional group is the general public. The fact that a certain technology is in principle allowed and the extent to which it is regulated is a public matter. This is however rarely part of the public debate as all the attention, if any, tends to be around specific technical options rather than the much bigger question of whether it is beneficial to have such technologies available in the first place.

There is no doubt there is some overlap across phases and stakeholders, but each phase offers a unique set of ethically relevant questions depending on the type of stakeholder involved.

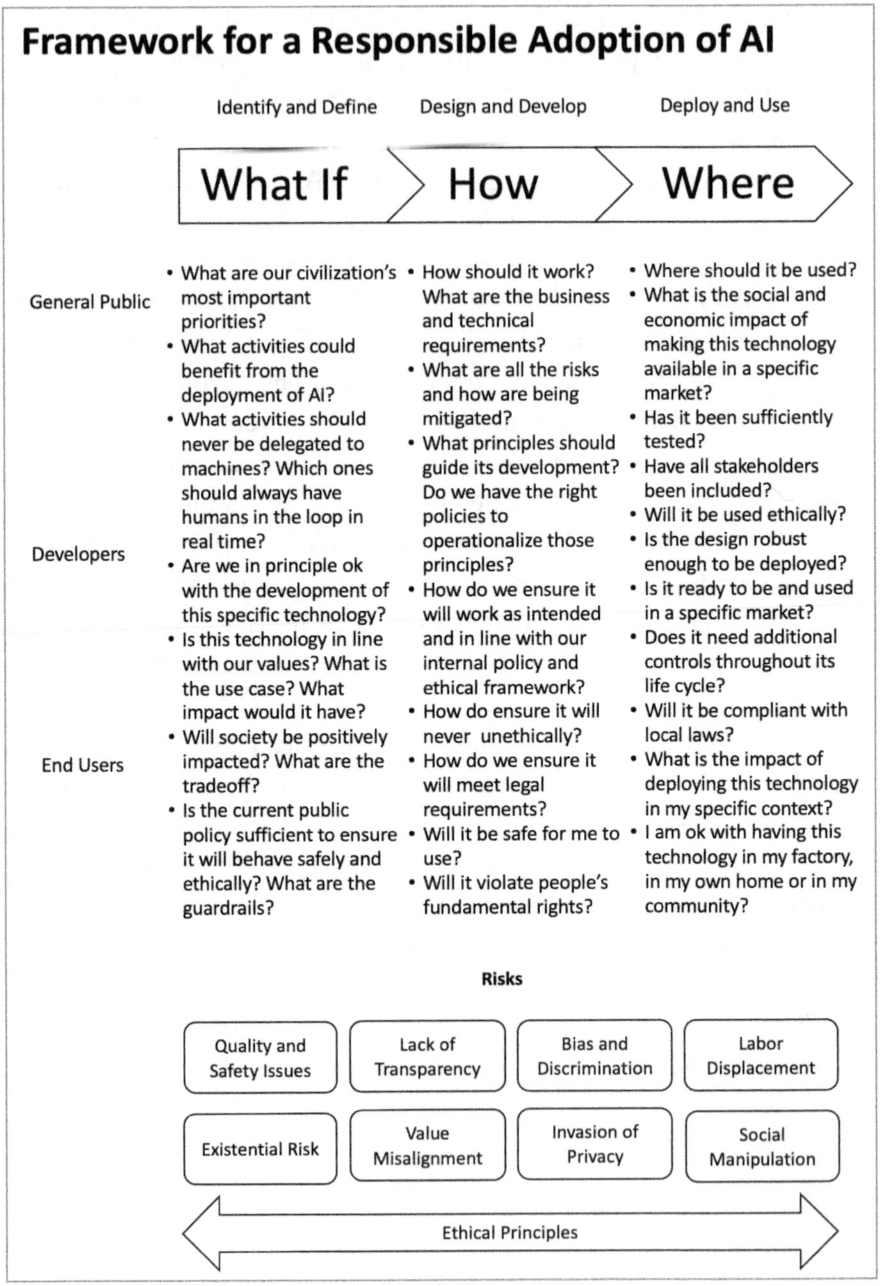
Framework for a Responsible Adoption of AI
Identify and Define
Design and Develop
Deploy and Use
What If
How
Where
General Public
Developers
End Users
• What are our civilization's most important priorities?
• What activities could benefit from the deployment of AI?
• What activities should never be delegated to machines? Which ones should always have humans in the loop in real time?
• Are we in principle ok with the development of this specific technology?
• Is this technology in line with our values? What is the use case? What impact would it have?
• Will society be positively impacted? What are the tradeoff?
• Is the current public policy sufficient to ensure it will behave safely and ethically? What are the guardrails?
• How should it work? What are the business and technical requirements?
• What are all the risks and how are being mitigated?
• What principles should guide its development? Do we have the right policies to operationalize those principles?
• How do we ensure it will work as intended and in line with our internal policy and ethical framework?
• How do ensure it will never unethically?
• How do we ensure it will meet legal requirements?
• Will it be safe for me to use?
• Will it violate people's fundamental rights?
• Where should it be used?
• What is the social and economic impact of making this technology available in a specific market?
• Has it been sufficiently tested?
• Have all stakeholders been included?
• Will it be used ethically?
• Is the design robust enough to be deployed?
• Is it ready to be and used in a specific market?
• Does it need additional controls throughout its life cycle?
• Will it be compliant with local laws?
• What is the impact of deploying this technology in my specific context?
• I am ok with having this technology in my factory, in my own home or in my community?
Risks
Quality and Safety Issues
Lack of Transparency
Bias and Discrimination
Labor Displacement
Existential Risk
Value Misalignment
Invasion of Privacy
Social Manipulation
Ethical Principles

## The WHAT IF Phase

The first phase, Identify and Define, or the WHAT IF phase, calls for an assessment of society's priorities and ethical implications of a new technology agnostic of how it will work and where.

Most of the attention around ethics is generally focused on the how: How can we ensure algorithms are not biased? How can we ensure our virtual assistants do not violate our privacy? How do we ensure social media does not spread false information? How do we ensure self-driving cars do not kill innocent pedestrians? That's important, but there are upstream "what" questions that we are rarely asking: What are our values? What is the best scenario for humanity? What society do we want to build? Would the creation of a specific technology be desirable from an ethical perspective? If so, under what conditions? The WHAT IF phase is aimed at addressing those questions across all its stakeholders.

For the general public this is really the phase in which there is the opportunity to define the right public policy. The analysis should be conducted at many different levels, ranging from fundamental questions about society's values and priorities, to more practical considerations around government incentives and oversight.

Some of the questions to be addressed in this phase are big-picture questions agnostic of a specific type of technological solution. Some questions will be focused on principles and values. What problems are we solving for? What are our key objectives and what are the metrics we will use to measure success? What are the principles that should never be violated? Some other questions will be around governance, oversight, and accountability: What oversight needs to be in place to ensure adherence? What are the roles and responsibilities of all key players? Is there clear accountability? What level of safety should be expected when introducing a new technology?

Some other questions will be related to the opportunity of employing machines in the first place and the types of presence we want AI to have in the long run. Key questions will include: What sectors need a stronger AI presence? What can be done to accelerate the deployment of AI in those areas? What activities should or should never be delegated to machines? What activities should always have humans in the loop in real time? What mix of artificial and human intelligence best promotes individual and social goals and values? For example, activities resulting in life and death choices, might be considered never acceptable unless there were humans actively in the loop. In this scenario, fully autonomous killer drones or certain types of fully autonomous vehicles would not be allowed. In many situations, however, rather than a yes or no answer, there will likely be some sort of conditional yes, where the technology is allowed only under certain conditions, in specific contexts, and with specific controls in place, setting the foundations for the HOW and WHERE phases. This is also the phase in which society should take the time to assess the opportunity cost of channeling resources toward certain technologies versus other priorities that will not be pursued because of that. For example, when we think about the development of virtual realities, not only do we need to evaluate the implications of having it around (such as using limited resources to build fake worlds), but also consider all the different initiatives or programs that could have been pursued to improve the human experience that will not get the resources. For instance, while we create virtual worlds where our avatars can experience beautiful simulations, the health of our biosphere (and its inhabitants such as ourselves) continues to pay the price. Since it is unreasonable to expect firms profiting from the creation of virtual worlds to stop doing what they are doing just because it may not be ethically acceptable, those types of assessments need to be made by the public so that the right incentives and disincentives can be put in place toward a more ethical use of resources.

Some additional questions may be specific to a new technology being considered: Are we in principle okay with developing this new technology (regardless of how it is going to work)? Is the current public policy sufficient to ensure it will behave safely and ethically? To what extent should it be regulated? What effects will it have on society? What are the risks and how will they be mitigated? Do the risks outweigh the benefits? What is the line that it should never cross? What are the guardrails? When assessing a new technology, the analysis should start by evaluating whether it would be in line with key principles and values, such as fairness, privacy, etc. If there are conflicting values, it will be necessary to assess which ones will take priority over others. For example, the idea of a new type of AI-enabled body scanner able to increase security at the airport might clash with the public expectations around personal privacy. Such assessments would also need to compare a proposed technology against existing policies to determine if a new technology would support or undermine strategic efforts. Assessing this early in the process rather than later allows more timely decisions on what type of proposed technology should be stopped, regulated, or even incentivized. For example, if fighting climate change was a top priority, the outcome of this type of assessment would be the ban of heavily polluting solutions and the promotion of technologies that could help reduce harmful emissions. This is not an easy task as it requires a serious analysis of the expected short-term and long-term impacts of a proposed technology on individuals, specific groups, communities, the environment, etc. In certain situations, the public might decide to forbid specific activities altogether. For example, with regard to the testing of self-driving cars, the general public might not be in favor of having prototypes of autonomous vehicles on public streets given the danger that they might cause to unsuspecting bystanders. This seems quite straightforward, but in 2018 when a pedestrian was killed by an experimental self-driving car by Uber in Arizona, it came to light that the Arizona's governor kept the self-driving car program a secret and citizens were never informed about the increased risk on the streets as a result of the decision to use public streets as testing ground. In other situations, the public might decide to do the opposite and proactively mandate the deployment of a new type of technology. For example, the public might decide to require car manufacturers to provide safety features that alert drivers that a pedestrian is crossing a street, or forbid manufacturers from allowing drivers to turn off such warnings.

Another big and often overlooked question is around the modification of the physical environment. To what extent should we alter our physical environment to address limitations of a new technology? When automobiles were introduced as the main means of personal transportation, we transformed our neighborhoods and our landscape by covering them with roads, highways, and parking lots to address cars' physical limitations. Are we going to do the same with AI? For instance, if we had a new technology that could improve highway safety but that also would require a major upgrade to the existing highway system, in what phase if not the WHAT IF one would we have the opportunity to collectively access pros and cons? When the first few prototypes of cars' combustion engines were developed over a century ago, we did not go through the exercise of assessing the impact of introducing them into our society. Only now we are realizing that proper controls should have been put in place to reduce cars' detrimental impact on our society from a social, cultural, financial, and environmental impact and only now are we starting to address some of those risks by introducing programs such as emission and fuel efficiency standards, mandatory safety features, etc. With the newest types of cars, such as electric self-driving cars, we cannot make the same mistake and automatically welcome the new technology without first assessing implications and opportunity costs. Even if proven to be safer than traditional cars, do they represent the best solution to address our transportation needs or are they perhaps just an improved version of a wrong type of solution? Would we be better off by focusing on more efficient means of transportation such as urban transit systems and regional high-speed rails instead?

Pursuing technology just for the sake of it would not be wise. A new technology may very well be part of the solution, but we first need to identify our objectives and then assess if we can keep the associated risks within our tolerance level. Only this way can we truly assess tradeoffs and opportunity costs. We need a serious discussion about what we want and what we do not want, regardless of the HOW. Do we want to ensure all humans have access to drinkable water? Do we want to stop the loss of biodiversity? Do we want humans to be free from violent dictatorships? Those are the questions that need to be asked. Technology can then do a lot for us to help us get there, but we need to start with the WHAT IF, without reducing the discussion to just the HOW.

For developers and other technologists involved in the ideation process of a new technology, the WHAT IF phase has similar types of questions asked by the public, but the analysis here is based on personal and corporate values: What is the use case? Is it in line with our corporate strategy and values? What principles should guide its development? Do we have the right policies to operationalize those principles? What impact would it have on our stakeholders? Do the risks outweigh the benefits? This is also when firms can first identify areas of concern and policy gaps either within their own organization or with regard to public policy, and develop strategies to address such gaps. For instance, a firm might realize the need to develop a code of ethics or to partner with other players in the industry to set up common principles and lobby for the development of legislation setting overarching standards.

Developers tend to overlook the WHAT IF phase and to focus more on the HOW. For example, if we think about the harm caused by social media to individuals and our society, it is impossible not to wonder whether those companies actually took the time to assess the implications of developing new algorithms aimed at increasing revenue designed to intentionally manipulate their users to change their moods or increase outrage. Had they taken the time to do so, would they have gone ahead with the design of such technology? On a similar note, when social media decided to add functions such as a like button, or a share button, or a retweet button, did they assess the implications of introducing that change? While the initial introduction of those new functions was barely noticed by the public, they had disastrous effects on people, especially the most vulnerable ones, and on our society triggering spikes of false information, bullying, depression, suicide rates, etc. By capitalizing on the outrage created by people with extreme views, those models have negatively affected individuals, communities, and even the foundations of our democracy.

Even the transportation sector has seen relevant examples of failures to identify ahead of time the unintended impact of a new technology. For example, when Uber decided to introduce a way for drivers to rate customers, they did not consider situations in which drivers might be inclined to provide a lower score to people with disabilities as they are generally slower getting in and out of the car. In the end, that's exactly what happened, and Uber found itself in a PR nightmare. A similar example of a lack of proper WHAT IF assessment, is with Uber's pricing algorithm. It was designed without considering situations in which high demand, resulting in exorbitant rates, could be associated with public safety needs, such as for instance the need to escape from a life-threatening situation. As reported in the news, complaints were filed by customers forced to pay exorbitant rates to flee a bombing in New York City.

For prospective end-users, the WHAT IF phase is mostly about assessing their appetite to delegate process steps to machines when such technology becomes available. When a company thinks about an emerging technology such as AI, the company should not only ask, How can we use it? but also, Should we use it? For example, this is the phase in which an HR function or a college admissions office would evaluate pros and cons to potentially automating a portion of their selection process by delegating to an AI model the scanning of applications. Another example would be a hospital's analysis of the opportunity to delegate to AI the triaging of patients admitted to the Emergency Room.

Across all stakeholders, the WHAT IF phase is fundamental for a number of reasons.

First, it puts the public in the driver's seat by assessing whether a solution is in theory desirable regardless of how and where it might be deployed. Many harmful or disruptive changes are presented to the public as unavoidable, but it is rarely the case. It is a public policy choice that should be addressed through the democratic process. When gambling devices giving out cash equivalent coupons were put inside many bars and pubs, it took a while to realize that those machines were creating a huge addiction and that families were suffering, and many governments have finally started banning them. The same type of assessment from the public needs to be done with all other kinds of technologies, including AI. Lack of attention from the public would inevitably create a legislative void that could be easily filled by the interests of a few players and harm large components of our communities. If we are worried about machines being in charge of our lives and want to ensure that we are the ones in charge, we should start having more WHAT IF types of discussions and finally exercise the same freedom we are so concerned machines will take from us. When social media was introduced 20 years ago, we did not go through that process and we have been paying a huge price for it (suicides, manipulation, disinformation, harming democracy, etc.). Even when we look at big societal changes such as labor displacements, we are not taking the time to talk about the WHAT IF. If that's not the type of world we want to live in, we are then shooting ourselves in the foot. Technology is a shiny object. It is very powerful because it allows us to get more with less. Rather than making it work for us we now often work for it. We tend to automatically assume that all innovation is worth pursuing and to chase technological innovations even if we do not know if they are going to bring any benefit for us. If there are negative connotations, they are usually perceived as technical issues that will be addressed with technical solutions later in the development and deployment stage. Public policy has been notably missing with respect to AI. It will be indispensable to influence the evolution of technology by creating the right set of incentives to channel efforts toward the right investments and the right controls to halt harmful applications. For example, given self-driving cars' abilities to help reduce the number of fatalities, the public might step in to expedite the development of such technologies. This is also when we would look at the implications of such technology from the perspective of those who make a living in the transportation sector that would be negatively affected by the introduction of self-driving vehicles. On the other hand, should the public be concerned about the impact of automation on jobs or any other risks, this is the phase in which such questions can be proactively addressed. Similarly, if the public considers it unethical to ever have an algorithm decide who gets selected for a job interview, the WHAT IF phase is the right place to start such a conversation, rather than waiting until the technology is developed, marketed, and deployed on people.

Second, the WHAT IF phase helps identify and address ethical questions early in the process so that investments can be channeled toward programs vetted by the public before the technology is developed. While it is not feasible to fully understand in such an early stage all the different types of harms that a technology might bring, it is not responsible to skip that analysis altogether. Given some of the risks involved the public might not have the luxury to postpone the debate as the consequences might end up being permanent. Additionally, while firms automatically tend to disfavor regulations, it is in the industry's own interest to have clear rules early in the development process as it promotes competition while also helping create trust among the public.

Third, it prevents situations in which we find ourselves throwing technical solutions at technologies that should not have been allowed in the first place. Given the lack of engagement of the public on big picture questions around AI ethics, the discussion has naturally been mostly led by technology firms and primarily focused on the technical options available to the developers building the technology. This has somewhat benefited technology companies as the conversation has been on the details of specific technical solutions rather than on the opportunity to allow those technologies in the first place. For instance, when looking at facial recognition penalizing certain racial groups, it is in the developers' interest to focus the discussion on how to avoid an unfair treatment of a group compared to other groups rather than answering the question about whether that technology should exist in the first place. If you are in the business of building and selling those technologies, you will understandably focus on the HOW, and rarely on the WHAT IF. This is not happening just in regard to AI, as it is part of a larger cultural shift toward techno-centric views according to which all of society's issues are just technical issues, rather than ethical, political, etc. A good portion of the political debate has in fact already been replaced by purely economic discussions, and such economic discussions are progressively becoming mostly a technical discussion to a point that the evolution of technology is perceived as the goal itself rather than a tool to improve our lives. As a result, we tend to think that there is a technical solution for everything, and if there is an unethical component of a model, we can correct it by changing the model. The reality is that there are areas where technologies and applications are unethical no matter how they work.

## The HOW Phase

The second phase, Design and Develop, is the world of HOW. Once it is determined that a technology could be beneficial, then the questions are all about how it should actually work so that it is safe and in line with our values. This is the phase where the majority of the debate around AI ethics has been so far.

In this phase the more active role will be played by the developers building the technology.

Some of the analysis will be on the intended outcome. How should it work? What are the objectives? How will this tool be used? What are the business and technical requirements? What are the intended outcomes? For instance, if we think about the development of an autonomous vehicle, this is the time in which we define the technical characteristics of the vehicle based on how it would be used. For example, will it be fully self-driving, or will it need some type of ongoing supervision from its passengers? What components will be customizable by its users? For example, will users be allowed to set up maximum speed beyond the speed limit? How should the car be programmed to respond in case of a mechanical failure or accident requiring a rapid choice between different scenarios? Will users be allowed to change safety settings to customize the way the car responds based on their own ethical views?

Some assessments will be completed to ensure compliance with internal policies and the law. Are there areas in which the machine might violate existing policies and legal requirements? For example, if we think about the development of an autonomous vehicle, this is when we would consider how to ensure the vehicle will meet government standards such as highway safety, environmental standards, etc.

Some questions will be around situations in which the machine might harm people. What are the risks? What are the possible unintended uses and unintended outcomes? What are the controls in place to mitigate the risk? How do we know it will be safe and work as intended? How will the model be tested? This is the phase where there should be an assessment around all the risks outlined so far, such as the ones associated with bias, privacy, transparency, safety, etc. Addressing them will be a collaborative effort across stakeholders with in-depth impact assessments, rigorous testing, and ongoing oversight across different levels.

As AI's performance heavily depends on choices made around the data used, discussions on the data sets to be used will surely be key in the discussions on the HOW. What data should the model be trained on and use on an ongoing basis? Will existing data be used, or will new data be sourced ad-hoc? How should the data be collected and processed? How transparent and auditable will it be? Will some type of data refinement or sanitization be needed? Given the fact that a lot of data is readily available, it is a common mistake not to take the time to assess the context and methodology in which it was collected as it might be unfit for the model being developed. The data reflects the goals and culture of those that collected it, so the same data set might be fine in certain contexts but unethical in others and there might be a need to proactively eliminate certain components. This requires a serious assessment of the available data to determine if sourced data can be used or if a new set of data should be collected.

In addition to the analysis around the type of data being sourced and used, serious consideration needs to be given to the risks associated with the way the machine processes it. For example, developers should proactively ensure that the model works as intended and does not cause harmful biases against specific groups either directly or via data proxies.

As part of the assessment there will necessarily be decisions to be made around tradeoffs. For example, one of the tradeoffs will be between accuracy and fairness. The current debate on ethical algorithms is about finding a balance between the need for accurate predictions and the need to not unfairly penalize specific groups. In fact, making algorithms more ethical often comes with a cost from an accuracy perspective. For example, if a dataset is used to assess your likelihood of repaying a loan, some sensitive data points such as gender or ethnicity might be intentionally removed to avoid the risk of unfair treatment of protected groups. By having less data, the prediction will likely be less accurate but also less likely to have biases negatively affecting those groups.

Other tradeoffs will be between privacy and public safety. On one hand we enjoy the freedom from surveillance or searches, but on the other hand we want to make sure that authorities do have access to the information they need to fight crimes, diseases, etc. With AI, this classic problem is even more relevant. For example, a phone application aimed at contact tracing between individuals exposed to infectious diseases for the safety of our communities, will surely be confronted with the need to protect the privacy of the users' medical record. As another example, considering that someone might confidentially tell a chatbot that he or she is planning to inflict serious harm to themselves or others, would there be an expectation for the chatbot company to monitor and report that to the authorities for public safety considerations, or should that be always protected as private?

Other tradeoffs will be around autonomy vs. control. Different firms might come up with completely different approaches. For example, just looking at self-driving cars, there are companies such as Google that have clearly endorsed the idea of having fully autonomous vehicles with no pedals or steering wheels humans could operate. This approach on one hand allows us to avoid the need of having humans actively in the loop ready to take control as necessary, but on the other hand offers no safety options in situations in which machines might not be able to respond safely to the environment. Other companies, such as Tesla, have given up altogether on having fully autonomous vehicles based on the assumption that AI will not be able to always ensure the safety of the passengers. This, however, does not address the fact that it might be unreasonable to expect humans not engaged in active driving to successfully take over at any moment without proper warnings. It is also likely that the driver might decide to activate the self-driving option not just for the convenience of it, but because unable to drive safely due to fatigue, distractions, intoxication, etc.

Finally, the developers building the technology can play an essential role as whistleblowers to bring to light unethical practices happening behind the scenes and seek further internal and external review as they are uniquely positioned to identify questionable policies and practices.

General Public and End Users might be less directly involved in this phase but still have a role to play. For example, citizens might want to advocate for having safety and quality standards in place to ensure products are ready to be deployed and introduce ways to monitor the development phase of technology either directly or via independent third parties. With regard to sustainability concerns, citizens might want to address items related to the environmental impact of the machines being built throughout their life cycle, from the sustainability of its supply chain to the recyclability of those materials.

## The WHERE Phase

The third and final phase, Deploy and Use, is the world of WHERE. Once the idea of a new technology has been vetted by the public, fully designed, and successfully tested, it finally becomes available to be used in real life scenarios.

For the general public this is the phase in which the need for any additional controls or guardrails may be assessed, to address any type of harm that was not fully addressed in previous phases. Having more details about the HOW will give the general public a better understanding of the technology's effect on individuals and society, including socioeconomic and environmental impacts.

For developers, this is the time to make any final assessment about the implications of making the technology available to the public. Is the technology ready for the market? Have we sufficiently addressed all the risks? How will we ensure it will behave as intended? For example, should a firm be ready to start selling a new software or perhaps a robot, this would be the opportunity to make their final go/no-go decision. The decision not to deploy might come, for instance, due to safety concerns for the technology itself, or perhaps about the risk of having such technology fall in the hands of ill-intended individuals.

For end users, it is the phase in which it should be assessed whether that specific technology should start being used in a specific context. For example, a firm seeking to automate the operations of a warehouse will need to assess how that would affect its employees, customers, local community, and all other stakeholders. For example, the firm might decide not to take advantage of the savings promised by the AI due to safety concerns or public relations risks. Something as straightforward as automating checkout lanes in a store or automating the collection of garbage bins on a city street might in fact come with a number of ethical considerations to address.

For consumers, this will be the moment to assess whether to adopt the technology in their lives. For example, placing a virtual assistant in my home might come with serious questions given the privacy risks associated with that technology. Some of the decisions might even turn into true ethical dilemmas. For instance, If I am trying to educate the public about deepfake videos, is it ethical to do so by creating and widely spreading them to help people realize how deceiving they could be?

The more customization options will be offered to users, the more relevant this phase will be going forward. For example, if people riding in self-driving cars are allowed to customize settings to determine the criteria used by the machine to assess how to behave in case of an upcoming accident, there will be an unprecedented opportunity to make life and death decisions ahead of an actual accident. For instance, particularly selfless users might decide to set it up so that the vehicle never hurts innocent bystanders even if that implies potentially serious harm to those inside the car. At the same time, someone else might decide to set it up so that the safety of people inside the car takes precedence over the safety of everyone else, including innocent bystanders.

## It will take a village

As we have seen, ensuring a responsible adoption of AI will take a village. Individuals, firms, activists, and governments will need to work together and across the whole technology's lifecycle. As a civilization we tend to be very reactive. With AI we will need to be super-proactive and take action on multiple fronts.

No matter what our role is (end-users, developers, etc.), we all need to think about both the small and the big picture. It is not reasonable to expect firms to stop doing what is beneficial to them unless there are specific disincentives. We need strong coordination across all phases of AI's lifecycle and all stakeholders, starting with a more focused public policy. Once you start looking across the whole spectrum, from the

WHAT IF to the WHERE, it becomes evident that there are a lot of ethically charged decisions to be made either explicitly or implicitly by a variety of players. Even before technologies are actually used, whether a gun or an algorithm, they are never ethically neutral.

We also need to ensure everyone feels responsible for the ethical implications of the decisions made across all phases. Maps such as the proposed framework will hopefully offer a tool for everyone to start thinking about the types of questions that need to be asked across the technology's lifecycle. When reflecting on the different phases of the framework, a developer working on improving HOW autonomous robots work, might become more engaged in big picture WHAT IF discussions on their intended use, and consider WHERE they might end up being used, their impact, and even possible unintended consequences, such as the risk of falling in the wrong hands. If, for instance, it turns out that the robot is being created to help with a military peacekeeping program, the developer should feel compelled to think about potential negative implications of having such technology around in the first place considering that it might end up being used by authoritative regimes.

Similarly, a citizen concerned about HOW facial recognition software might negatively affect certain groups, might start questioning why it is appropriate to have such technology in the first place, demanding a serious assessment back in the WHAT IF phase so that it can be addressed at a public policy level. With so many questions to be answered, it will take a lot of time and a lot of work to get to the right solutions, so the earlier we start the better it will be for everyone.

## CHAPTER'S KEY POINTS

- Rather than freeing us from our responsibility, AI actually multiplies it as it magnifies our power.
- Most of the discussions around ethical AI focus on how the technology works, but they fail to adequately assess whether

such technologies should exist in the first place and if the resources needed to develop them should be used in support of other more pressing efforts.

- The proposed framework offers a map to navigate ethical considerations across the AI's life cycle providing examples of key questions for the general public, developers, and end-users.
- Ensuring a responsible adoption of AI will take a village. Individuals, firms, activists, and governments will need to work together across all phases of the technology life cycle.

# International Cooperation

As with many other global challenges, such as the current environmental crisis or human rights violations, ensuring an ethical use of AI will require a village, the global village. Any meaningful action will need to be taken at an international level because AI is already ubiquitous, and it feeds on data that can easily travel across the globe. The old saying "my house, my rules" does not apply when dealing with AI. It will also need to be a collaborative effort because it is not reasonable to expect individual countries to voluntarily give up the power that comes from AI unless it is part of a coordinated global initiative.

After the end of the second world war, the planet came together to define a set of fundamental values, later adopted as international law in the United Nations Universal Declaration of Human Rights. The same type of effort needs to be made in regard to AI to define commonly accepted minimum standards and introduce legally binding rules with a strong enforcement and oversight framework.

One of the challenges will be to figure out the correct level of oversight across international, regional, and local efforts. Additionally, it is not going to be easy to find an agreement among rival nations or with countries already concerned about being left behind. For that reason, we should begin immediately and focus on the most obvious risks and go from there. For example, it would be a great start if we could get everyone to agree that AI cannot be deployed for things such as the automated launch of nuclear warheads.

At a regional level we are starting to see international efforts to address different areas of risk. For instance, the European Union has taken some important legislative steps to address the use of AI in high-risk applications, and to protect the public against privacy violations, online violence, racial discrimination, and fake news.

Somewhat in parallel, there have also been a number of efforts aimed at better leveraging existing international human rights frameworks such as the United Nations Universal Declaration of Human Rights. Such efforts are headed in the right direction for two main reasons. First, because despite being written decades ago, those principles are wide enough to provide coverage across many issues related to AI. Second, they have already been adopted as binding international law, and even replicated in some countries' own constitutions. The challenge, however, is the lack of an enforcement framework that allows individuals and communities to have such rights protected.

### Call for the international ban of autonomous lethal weapons

One area where experts and the general public have been particularly active is around the topic of autonomous lethal weapons.

Thanks to recent advancements in technologies such as autonomous navigation, computer visions, facial recognition, and cooperative autonomy or swarming, AI can now support a variety of military applications, from autonomous tanks to mini killer drones able to autonomously identify, target, and eliminate threats, including humans. This scenario, that until recently only belonged to the realm of science fiction, has caused great concerns among experts and the general public.

Just as an example of those concerns and many leading experts' strong positions, in 2015 nearly 4,000 AI and robotics researchers called for a ban on lethal autonomous weapons (LAWs) by mandating human control. The scope of the proposed ban was specifically for where humans are not in the decision-making loop in real time. In 2017, 137 CEOs of AI companies asked the UN to ban LAWs. And In 2018, 240 AI-related organizations and nearly 3,100 individuals took that call a step further and pledged not to be involved in the development of LAWs. If successful it would be a step in the right direction even if very limited in scope, as the ban would not outlaw drone warfare or other semi-autonomous AI-powered weapons where humans are in the loop. It would also not outlaw autonomous missile defense systems as long as they are designed to defend against incoming missiles rather than targeting humans.

It is a defining moment, but the time window for addressing it is getting smaller and smaller every day. Without a ban, a global arms race would be virtually inevitable. All you need is just one military power to start pursuing that technology and we will soon find ourselves in the midst of an AI arms race, which will lead to cheaper, deadlier, and more ubiquitous autonomous weapons.

Establishing legally binding agreements will be tough. Enforcing such agreements will be even tougher, especially when dealing with the interests of the strongest military powers. With nuclear weapons, international negotiations were relatively simple as they were all around the number of missiles, inspections, etc. With AI, if a nation feels like they have a new, powerful discovery, it will be easy to keep it secret and they will not have much of an incentive to disclose it to others.

On a more encouraging note, this is not particularly different from the efforts banning bio-chemical weapons. By creating enforcement mechanisms and a strong public stigma on the use of such weapons we have not seen large-scale violations of these bans despite being relatively easy and affordable to make.

To be clear, most of the scenarios that people are worried about are already prohibited under the laws of war or international humanitarian law, such as the United Nations Convention on Conventional Weapons (CCW). Those binding principles lay out a set of parameters for what is acceptable and what is not acceptable in any weapon system, including autonomous weapons.

One example is the principle of distinction: military forces cannot intentionally target civilians. They can only intentionally target other military forces. Weapons cannot be used or just built unless they are able to comply with this principle. The same principle also applies to similar situations, preventing, for example, the use of force against those who have surrendered. Another key concept is the principle of proportionality: collateral damage that affects civilians or civilian infrastructure cannot be disproportionate to the military necessity of the target that is being attacked. Additional conventions often mentioned as already somewhat banning autonomous weapons include the Geneva Convention, the Marten's Clause, and the International Covenant on Civil and Political Rights. However, most experts believe that the international law principles are dramatically insufficient and that we need to go beyond those principles and introduce a direct ban on LAWs expressly forbidding all situations where humans are not in the loop of decision making in real time. Such a treaty would also help clarify what are acceptable uses of AI with respect to lethal decisions and what aren't.

Drawing clear lines around the acceptable and unacceptable uses of this technology, including whether we allow AI to take human life, will be essential for establishing international collaboration and governance around LAWs. Such regulation efforts or lack thereof will also shape the kinds of weapon technologies that proliferate in the 21st century. Many countries have shown support for recent efforts led by the United Nations, but there is still strong opposition from some of the biggest countries. Given what is at stake, everyone's support will be key.

There is no solution in sight besides the one that can be achieved via cooperation. Cooperation is humanity's superpower. This is how we have managed to accomplish the best things and that's how we will address the challenges and opportunities that come with AI.

## CHAPTER'S KEY POINTS

- Like many other global challenges, the risks and ethical challenges that come with AI cannot be fully addressed unless they are tackled at a global scale.
- Concerns about lethal autonomous weapons and the risk of an arms race have triggered encouraging initiatives seeking a global ban of unsupervised military applications.

# Legislation and Oversight

Letting a fox guard the henhouse?

Legislation and government oversight have been minimal so far. The general idea since the nineties has been let's start playing now and we will figure out the rules later. The hope was that firms would self-regulate, creating their own rules and introducing effective mechanisms to operationalize them.

Decades later, most of the biggest players still lack a basic framework to address ethical issues related to AI. Countless scandals such as the 2016 Cambridge Analytica disaster have confirmed over and over the failure of any strategy solely based on self-regulation. Even after such failures, the industry has been very slow in its response. For example, it took Facebook 4 years after the Cambridge Analytica case to put together an Oversight Board, a quite modest effort that has not yet demonstrated to be up to the task. Another example of a disastrous attempt to self-regulate was Google's Advanced Technology External Advisory Council, created to provide guidance on how to ethically develop new technologies such as AI. The Council was decommissioned just a week after its birth due to protests around the adequacy of some of the members selected for the Council.

The reality is that historically, across virtually all sectors, it has never been a good idea to have a fox guarding the henhouse. Firms do not have any incentive to deviate from what makes sense for their bottom-line. It is not reasonable to expect that for-profit firms will spontaneously give up profits for just ethical reasons unless the decision is driven by other factors such as customer needs, compliance requirements, or reputational risk considerations. Firms do a good job assessing acceptable levels of risk for the public only when firms' interests are very closely aligned with those of the public at large, which is rarely the case. We cannot rely on the goodwill of companies to protect us from abuses caused by technologies in which they have a vested interest in developing and deploying. The free market is excellent at creating wealth, but not at planning for the good of society because it is not set up to consider externalities such as its impact on people and the environment. A lack of rules generally results in lowering the bar, hurting the most virtuous ones. Once a player decides to lower its ethical standards to gain a business advantage (less oversight, more sales, etc.), this can easily trigger a chain effect across competitors just to be able to remain competitive (first offenders' advantage).

The risks that come with AI require a serious effort to develop a legal framework, and a plan to enforce it. Code of ethics, good will, good intentions, and communication campaigns outlining moral principles are certainly a good start, but they are not sufficient to ensure technology works well for the common good. The strongest means to enforce the collective will is the government's ability to develop and enforce laws.

## Key areas of focus

Legislation and oversight can help with a number of objectives. First, public policy can help guide the deployment of AI toward the type of human activities that represent a good fit with AI and limit adoption in areas in which it would be detrimental to our society. Rather than blindly welcoming AI in every sector and across all tasks at the risk of foolishly liquidating the best humans can offer, public policy can help prioritize efforts so that we deploy AI where we can also get more in return in the long run. Public policy can also help ensure that a person is meaningfully in the loop for all the most critical activities that might be delegated to AI, from health care to the military.

Second, legislative action can help protect social values and moral principles by introducing clear rules and ensuring responsible parties are held accountable. An area in which legislation can greatly help is in regard to autonomous vehicles and the accountability expectations in case of accidents. With the introduction of Generative AI, we also need to urgently revisit our current copyright laws to ensure that the use of AI-generated consent is done transparently (requiring, for instance, specific types of watermarks) and with fair attribution and compensation to the original creators. Even for applications such as social media, legislation can help hold firms accountable for the harm caused to their most vulnerable users.

Third, by introducing ongoing governance and oversight, regulations can be very effective at addressing public risk where the potential victim of a harmful technology is not in a position to directly stop the harm. Rather than solely relying on companies' high-level aspirations and their own risk-reward evaluations, governmental institutions can help ensure a fair distribution of risk. For example, to ensure sufficient levels of security and auditability of the technology, many are advocating for the introduction of the same type of rigor we see in other sectors such as air travel with AI-specific certifications, third party inspections, and independent assessments to ensure that models meet certain security standards, help identify the cause of harmful failures, and improve performance and safety over time.

Fourth, government oversight can help promote competition by addressing potential concentrations of power and help ensure a level playing field among players. The lack of action in this area has been particularly noticeable. For example, many have questioned the decision to allow Facebook's acquisitions of other dominant platforms such as WhatsApp and Instagram.

Finally, regulations can even help drive innovation and increase people's trust of those who develop them. Regulations have in fact already demonstrated their ability to create opportunities for new industries to flourish. For example, car safety regulations have opened the door to new markets for safety technologies and, by making cars safer, helped car manufacturers develop a bigger market. Regulations can also help create a more efficient market and create better conditions for innovation. For example, in the United States, digital medical records cannot be fully used for research purposes due to the highly fragmented legal landscape across states. A more active role of the government to harmonize regulations would open the door to new opportunities for medical and research purposes.

Many of the leaders in the AI industry have openly recognized the need to start regulating the industry. For example, in 2017 Tesla's CEO Elon Musk shared his belief that "…it is important to get the rules right. Like in sports, if the rules are not set up properly it is not going to be a good game". As another example, Microsoft's president, Brad Smith, while discussing facial recognition technologies called for "thoughtful government regulation and for the development of norms around acceptable uses. In a democratic republic there is no substitute for decision-making by our elected representatives regarding the issues that require the balancing of public safety with the essence of our democratic freedoms." The good news is that we do not have to start from zero. Many of the largest players have already set up policies and governance framework, to manage their technology, audit how inputs lead to outputs, and evaluate its impacts. By analyzing the best examples of such initiatives, it would be relatively easy to implement generally accepted AI standards and metrics as it has been successfully done in other areas such as the generally accepted accounting principles used by US firms to report on their financials. Independent regulatory bodies would then be in charge of overseeing compliance and auditing firms that develop and use AI.

Where, when, and how much?

While innovation is not easy, finding the right level of governance and oversight is much more difficult to achieve as it needs to find the right balance between protecting people from harm and promoting innovation, so that we can have a higher level of trust in new technologies without missing the opportunity for beneficial breakthroughs.

One challenge, for example, has to do with the timing of the government intervention and its potentially negative effect on innovation. How early in the development process should the government intervene to regulate? If regulations come early in the process, it is very inexpensive for governments to regulate and for the firms to comply, but it is also difficult for the government to know what risks will need to be mitigated. It could end up unnecessarily hurting innovation causing a chilling effect. If regulations come later in the process, it might be easier to address risks as it will be more clear what harm may be caused, but they might become very costly to comply with. Once unethical practices become widespread it is also more difficult to take action, as once the bar has been lowered the public tends to be less engaged. For example, social media regulations would have been more feasible if started 20 years ago, but it is also true that at that time we couldn't possibly conceive the types of risk that social media was about to introduce. The more a specific technology is part of our world, the more difficult it is to intervene. Just as it's impossible to truly pull the plug on the Internet, so too would it be impossible to truly pull the plug on a widely used AI model, so early identification of risks is key. The United States has so far mostly adopted a "wait and see" approach, with the exception of specific industries such as the financial sector, already under significant regulatory scrutiny with regard to the risks associated with the use of statistical models, including AI. The European Union has been much more proactive with initiatives such as the European Commission's General Data Protection Regulation and The Ethics Guidelines for Trustworthy Artificial Intelligence. The reality is that regulations tend to come late as the democratic process requires time to come up with the right rules and enforcement infrastructure. It also tends to be a quite reactive process where action is only taken after the fact as a result of public outcry. With AI we need to be proactive rather than reactive as we cannot risk getting there once it is too late. For example, if we look at the risk of having ongoing state surveillance, we only have a small window of time to take affirmative steps to prevent that from happening while we have the freedom to do so.

Another challenge is to find the right level of granularity. Regulations are more effective when very specific, but regulators can be very specific only when addressing issues that are well understood. Given the time needed to develop such rules and the speed of innovation, regulations can quickly become obsolete, especially if very specific. Rather than trying to have very detailed rules, we are much better off with high-level frameworks that work well across all types of technologies.

It is also difficult to assess what types of rules should be sector-specific and which ones should be applied horizontally across the board. On one hand, not every sector warrants the same level of scrutiny and oversight, as they expose citizens to significantly different types of risks. On the other hand, there are substantial risks, such as privacy risks, that require similar approaches across all sectors. Given the economic burden of regulations, the key is to find the right balance between horizontal and vertical regulations and specifically target high risk areas.

While many now recognize the need to actively regulate, governments are truly behind and unprepared to handle the challenges coming with AI and the gap is getting wider. It is an issue, as it is impossible to regulate what we do not understand. The best example of such a gap is the famous US Congressional Hearing on Facebook's business practices in 2020. The questions that the committee asked Facebook's CEO Mark Zuckerberg during his testimony were so anachronistic that it instantly became a viral social media meme.

More in general it is objectively difficult to get regulations right in a field in such a constant and rapid evolution. The general public is understandably not prepared to navigate this new and highly complex landscape. There is also an obvious information asymmetry between developers and the public as only developers and tech firms are in a position to really understand how the technology works. This incompatibility between technical complexity and the democratic process is not unique to AI, but particularly concerning given the power of such technology.

The best scenario is where innovation is incentivized but within clear guardrails, with a good balance between ethical needs and economic interest. Innovation is not a value that should be placed above all other ones. While we should rightfully celebrate innovation done well, we must be careful not to assume that any and all technological innovation is something we want. Our communities might very well decide to place values such as privacy and democracy above others such as the drive toward technological innovation and take legislative action accordingly. Interesting examples of laws enacted to limit the use of specific technologies in order to protect higher values have recently made their debut. A number of US cities and states have recently introduced bans on law enforcement agencies' use of facial recognition technologies. Other cities have banned cashless stores to prevent discrimination against citizens who do not have access to payment cards or apps and therefore cannot make purchases unless stores accept cash. Some other cities have banned altogether the use of short-term rental apps such as Airbnb in response to the perceived negative impact on the local real estate market.

## Addressing the inequality caused by AI

As we have seen, AI has brought amazing new opportunities across many sectors, but it has not been good news for everyone. While technological advancements help make the pie bigger, the wealth is not proportionally distributed as the capital tends to be unequally owned. What that means is that technology becomes a multiplier of wealth for those who control capitals, and a multiplier of poverty for those who don't, starting with low-skilled workers for whom it will get harder to find work. What can be done collectively to address these situations of inequality?

A number of ideas have been proposed to help manage the transition that can be deployed right away.

The general concept across the options being discussed is to allocate the benefits of the increased productivity brought by machines to different sectors of society, rather than just the most privileged groups. The entities causing innovation-driven externalities, such as poverty, inequality, and negative effects on the environment, would be held accountable to be part of the solution. The most popular proposal is the introduction of a universal basic income (UBI). According to such a proposal, the additional productivity created by machines would be redistributed to those left behind in the form of monetary distribution. Without such transfers, many argue income inequality will rise, and large parts of society will be increasingly penalized, generation after generation. Considering that AI research has been largely funded by the government through grants for universities and other public programs, the transfer of a part of the wealth generated thanks to AI to those who are in need would not only be just a subsidy but also a way to redistribute the resources generated thanks to an investment made by the public.

As many are uncomfortable funding programs that send out monetary contributions, as it will be difficult to control how the cash will be used, some have proposed different types of in-kind support such as direct coverage for essential needs like healthcare, education, and professional training. Programs where citizens are offered training to start a new career are particularly desirable not only for financial purposes but also as a way to promote self-esteem and a sense of belonging. As Voltaire stated, work frees us from three great evils: boredom, vice, and need. UBI only takes care of one of them: need. It is therefore important not to limit our focus on just the most material aspects of this challenge. The financial coverage to fund such programs could come from government tax initiatives aimed toward firms profiting from automation, which could be appropriately called a robo-tax. It is also true that such programs, by helping address issues exacerbated by inequality such as violence or substance abuse, might even end up paying for themselves.

Even unions can play an important role. For example, unions might propose the introduction of retain and retrain clauses for AI-driven organizational changes. For example, collective agreements could include provisions such as advance notice of layoffs, and/or free reskilling for new jobs created by the development of new technologies. Some other ways to redistribute the value provided by automation could include proposals such as the reduction of the work week to 30 hours or the introduction of other benefits such as vacation time and family leave.

It would be very unfortunate for our civilization if something as promising as AI would end up creating more poverty and more inequality while transferring more wealth to a few more billionaires. It is reasonable to expect a strong opposition by those who would benefit from those scenarios, but we need to make a choice. We need to find a way to make the AI genie work in our favor, in a way that is beneficial to the whole of humanity, ensuring that everyone can have a decent life in a decent society. The alternative, which no one wants, would be a breeding ground for anti-AI or no-robot protests, leading to even greater violence and civil unrest.

## CHAPTER'S KEY POINTS

- The private sector has failed to successfully self-regulate to address the risks related to AI.
- Government intervention can help ensure a more ethical AI while also promoting competition among all players, but it will not be easy to get the timing and scope right.
- AI will likely become a multiplier of poverty and inequality unless proper legislative measures are put in place to redistribute its benefits and address its impacts on communities and the environment.

# Self-Regulation and Industry Standards

Many firms have started to think seriously about the intersection of ethics and AI and to acknowledge that more needs to be done to ensure a deployment in line with companies' ethical principles and values. The drivers of those efforts are rarely based on purely ethical considerations, but also based on other priorities such as ensuring compliance with policies and legal requirements, adherence to social responsibility strategies, risk assessments, etc. By proactively addressing those issues, firms not only can reduce the risk of public relations disasters but also increase the public's trust in those technologies and reduce the need for government intervention.

## The Intersection between AI ethics and law

Both ethics and laws are constantly evolving and vary greatly across individuals, communities, and regions. Ethical norms such as values, rights, duties, and responsibilities, often end up becoming laws. For this reason, firms tend to consider their ethical responsibilities fulfilled just by not violating laws. The reality, however, is that solely relying on legal compliance can expose firms to substantial risk as there are scenarios for which the law might be vague or completely silent.

Firms will then still need to proactively address those scenarios to mitigate those risks.

There are also scenarios in which there is an actual mismatch between moral principles and the applicable law. This might be caused by a number of situations such as the rapid evolution of ethical views, delays in the legislative process, etc. In some scenarios, a firm might be asked to comply with a law that goes against the firm's ethical commitments and therefore it might decide to violate the law to stay true to its ethical views. A typical example is a global company that refuses to release its users' personal data to an authoritarian government persecuting political activists in fear that users would be persecuted. In other scenarios, the firm might refrain from doing something that is legal but against the firm's principles. For example, a technology company might decide not to support work for the development of LAWs for ethical considerations. In some scenarios the law might be in line with the firm's ethical view, but the firm decides to do even more than what is required by the law. An example would be a firm proving stronger privacy protection to their customers than the minimum required by the law which they deemed insufficient. This type of ethics in which firms do more than the minimum legal requirement is sometimes called Post-Compliance or Soft Ethics.

## The Intersection between AI Ethics and Corporate Social Responsibility

Firms' decisions might also be tied to specific corporate strategies such as Corporate Social Responsibility (CSR) programs or Environmental, Social, and Governance (ESG) programs, aimed at maximizing shareholder value by integrating social and environmental concerns in their business operations. Thanks to those programs, firms might end up pursuing more ethical options, even if they are not necessarily the outcome of moral considerations, but rather just in support of a larger business strategy to improve specific areas such as employee satisfaction or customer engagement. For example, a firm might come up with charitable programs in support of local communities as a part of the larger strategy to increase market shares in a certain area. While there is surely an ethical component, the biggest driver is tied to business interests and meeting shareholders expectations. This romantic idea of companies choosing selfless altruism only appeals to those who have not had much experience working in global companies.

## The Intersection between AI ethics and Risk Management

Identifying and addressing the risks that come with AI is not terribly different from what firms already do to address areas of risk such as legal, operational, and reputational risk. Risk management techniques can help firms with the identification of the likelihood and severity of events that can have a negative impact, determine tolerance levels, and develop plans to mitigate that risk to an acceptable level. Most of the decisions in which firms decide to take a strong ethical stand are not based exclusively on moral principles, but on their risk appetite, especially with regard to reputational risk. Firms will avoid practices perceived as immoral by their stakeholders to avoid a negative public reaction with unfavorable repercussions on their business. This is not a negative dynamic because the more firms understand the intersection between ethics and risk management (and its impact on their bottom-line), the more they will be inclined to create stronger programs on AI ethics.

By correctly identifying the risk of having certain programs perceived as ethically unacceptable, firms can deploy mitigation plans that not only help avoid costly public relation nightmares, but also contribute to the public's trust of new technologies such as AI. For example, after protests from its employees and other stakeholders, Google took a second look at one of its contracts with the military and decided to end its support for military programs using AI on the battlefield.

## Building an AI ethics framework

Despite the positive progress made as a result of the efforts described above, firms continue to be behind on the ethical side. Even the largest multinational tech companies admit they are still struggling in this area. This is in part due to companies' competitive pressure, but also due to the fact that the tech sector is just still relatively new to ethics. Professions such as lawyers or doctors have had ethical requirements for decades with professional organizations monitoring compliance (and sometimes revoking licenses), but that's not yet the case for data scientists or software engineers.

So, how can we go from discussion into action? In order to effectively self-regulate, firms need to make AI ethics a priority by putting together a cross-functional governance program with clear principles and operational processes throughout each stage of products' lifecycles. The first step is to identify the risks that need to be mitigated and then create a strong governance program built around people, policies, and processes.

### People

Firms need to ensure they have the right people at the table: not only qualified and experienced, but also personally passionate about ethics and committed to driving the right discussions about the design and use of AI. It is also key to ensure that teams have a high degree of diversity across multiple dimensions such as gender, race, ethnicity, age, educational background, etc.

Additionally, there must be clear roles and responsibilities for all people accountable for driving ethical outcomes. For instance, it might be appropriate to create ethics-specific roles such as Chief Responsible AI Officer or Chief AI Ethics Officer. Such leaders would be on point to define and coordinate the firm's approach to responsible AI, including the definition of the internal policies and the oversight of the governance process. Amid growing public pressure and mistrust, major firms such as Microsoft and Salesforce have already started creating these types of C-suite positions fully dedicated to AI ethics and policy to add ethical perspectives to the design, development, and deployment of AI systems. Given the wide range of skills and technical expertise needed to perform such a role (ethics, risk management, regulatory requirements, business acumen, etc.), it might be difficult to find qualified talent. It is therefore reasonable to expect that such a function will be delivered not by just one person but by a cross-functional team with experts from different sectors and functions.

In addition to AI ethics-specific roles, everyone at the firm should receive appropriate training around ethics, making it clear that identifying and escalating ethical concerns is everyone's job across all functions and levels of the organization.

### Code of Ethics and company policies

As a foundation for an ethical framework, firms should develop Codes of Ethics and company policies to capture core principles and values and define how ethical issues will be handled. Examples of principles commonly included in Codes of Ethics to support a responsible approach include transparency, fairness, accountability, inclusiveness, privacy, safety, etc. Those might also be complemented by more overarching company values such as respecting the dignity of individuals, protecting social values and public interest, sustainability, etc. Codes of Ethics would then work in conjunction with any additional policies such as AI policies, Data Management policies, etc.

The biggest challenge is to have principles that are specific enough to provide meaningful guidance on specific applications, while also being generic enough to allow room for future technologies.

As there might be cases in which principles come into conflict with each other, it is also advisable to create some type of prioritization methodology and define an approach for the interpretation of such principles where past decisions can be leveraged as some sort of precedent, as typical in some legal systems.

Another challenge, given the high variability of ethical expectations and laws across regions, is identifying a global approach that could be adopted by all countries. This might, however, be quite difficult as certain regions might have very different laws and ethical expectations.

## Processes and governance

In order to operationalize the policies, an ongoing governance framework with ad-hoc processes and routines needs to be put in place for the day-to-day management and oversight of an AI ethics program.

Processes need to be in place to assess initiatives across their lifecycle, from the initial identification of the opportunity to its development, deployment, and retirement. These steps should be followed not only for each technology but also for each product leveraging in-scope technologies. Creating such a review and vetting process should be assessed through an ethical lens, and should be built upon an ongoing tollgate process, becoming part of the technology program's workflow, leveraging standardized and auditable risk assessments and checklists.

An AI Ethics Board should also be in place to oversee the overall governance and serve as an escalation point. In order to be as effective as possible, it should have a wide range of talent and expertise across functions and disciplines, from ethicists and social scientists to developers and engineers. As technologies might be used in a variety of social contexts, such as an office, a home, or a public street, it is key to have good representation of different groups in order to better identify specific risks. It is also important to intentionally reduce information asymmetries across participants as some of them might be at a disadvantage with regard to access to data and/or technical matters. Finally, there should be an audit trail that shows the decisions made step by step as part of the development and implementation process to increase transparency to external parties.

### Industry standards

With AI there have been very interesting attempts to develop ethical standards with key industry players like Partnership on AI, Open AI, and the Future of Life Institute.

Industry associations have also taken meaningful first steps such as with the Institute of Electrical and Electronics Engineers' Global Initiative for Ethical Considerations in Artificial Intelligence and Autonomous Systems, which has a strong emphasis on protecting human rights and the principles of responsibility, transparency, and education.

While helpful in promoting the right type of debate, industry standards have intrinsic limitations. Committing to them is merely optional and, even when firms do so, lack of compliance is typically not legally binding. In absence of ad-hoc legislation, the hope is that industry standards might one day be looked at by courts as "duty of care" standards, making such benchmarks the minimum standards for those industries with the same mechanisms in place today in more mature industries such as the pharmaceutical and the transportation sectors.

## CHAPTER'S KEY POINTS

- Most companies' ethically relevant decisions are not made based on codes of ethics, but by the need to ensure compliance with policies and legal requirements, adherence to social responsibility strategies, and in response to risk considerations.
- To turn moral principles into action, firms need to proactively operationalize them with programs built around people, policies, and processes.

# Education and Research

## Education

Among the many risks associated with AI, one underlying risk across all areas is the overall lack of knowledge and experience. The biggest AI disasters so far have been caused by a mix of incompetence and ignorance. Even among the biggest experts in the field there is often very little understanding of foundational aspects of the many disciplines AI intersects with, including ethics.

We cannot reasonably expect an ethical implementation of AI unless all the technical experts involved are at least somewhat familiar with disciplines such as philosophy and applied ethics. Courses on AI ethics should be part of the standard curriculum for anyone else involved with the life cycle of AI from developers and data scientists to end users such as business managers, governments, and non-governmental organizations. This should not be just about learning theories and theoretical frameworks but also about developing a mind set and sense of responsibility for the ethical implications of new technologies. Data scientists and software developers, for example, rarely feel it is their job to consider those issues, as they are normally addressed by other functions such as legal and compliance departments. This will need to change. We cannot reasonably expect software developers to become impeccable ethicists, but we do need to help create an environment in which everyone feels responsible about the ethical implications of the technology being developed and its impact on individuals. Developers and organizations' leaders cannot delegate this responsibility to consultants or third parties. They need to be prepared and educate themselves on the ethical implications of the choices they make. Given the importance of these considerations, some people have even proposed the introduction of a Hippocratic Oath (like the doctors' oath) on ethical AI as an additional way to reiterate everyone's role.

Similarly, as more professionals, such as lawyers or doctors, rely on those technologies to make decisions, their professional curriculum needs to include foundational elements of emerging technologies such as AI. This will help them be better positioned to have a more active role helping to identify situations of risk. Given the rapidly evolving nature of the landscape, this should not be a one and done effort, but part of an ongoing education.

## AI Safety Research

As AI gets stronger and more widely applied, it becomes more critical to ensure resources are channeled toward AI research focusing on key areas such as safety, civil liberties, privacy, and human rights protection. Governments will need to take the lead on this and create ad-hoc councils and funding programs not only to incentivize and monitor the research work but also to democratize access to those resources and knowledge.

From a safety standpoint, the objective is similar to the ones in other fields of technology: creating AI that does not have bugs, does not malfunction, cannot be hacked, and so on. While it may be a big nuisance if your laptop crashes or gets hacked, it is even more important that it does not happen if an AI system controls your car, your airplane, your pacemaker, your automated trading system, or a critical infrastructure such as your power grid.

Over the long term, the focus should be around the existential risk that comes with AI. Technology's capabilities are growing exponentially and definitely faster than humans' cognitive abilities. Even if we address all the cultural problems and become the best humans we could possibly be and build the best society we could possibly build, we will still have a number of biological limitations that will put us at a disadvantage. If we want to continue to be the ones in control, we need to ensure that at any given time the strength of our controls is greater than the strength of the technology. This will require a serious collective investment in AI safety to ensure AI is in line with our values and goals. Something as simple as the requirement of an on/off switch and connectivity shut-offs for all machines, would greatly help mitigate the risk of scenarios in which a machine might resist our attempt to shut it off or to change its goals.

Our civilization will continue to flourish only as long as the growing power of technology does not surpass our ability to control it.

## Education as our first line of defense

More broadly, the most effective defense against the risks introduced by AI, from large-scale social manipulation to more extreme, systemic threats is education itself. The purpose of education is not to replace ignorance with certainty, but to replace ignorance with doubt: the ability to question the status quo, examine claims from different perspectives rather than accepting them at face value, and identify new paths to new solutions. In an environment saturated with automated content and persistent misinformation campaigns, citizens must remain active participants, capable of evaluating sources, questioning narratives, and recognizing uncertainty. Education, in this sense, is less about providing answers and more about cultivating the habit of skepticism that allows democratic societies to function in an AI-mediated world. Strengthening curricula around critical thinking is therefore essential to equip citizens with the intellectual tools needed to assess information, ask the right questions, surface the risks, and make informed decisions for themselves and their communities.

## CHAPTER'S KEY POINTS

- Players involved in the deployment of AI should have a better understanding of the ethical implications and feel responsible for its impact.
- Strong investments are needed on AI safety as our ability to control AI must evolve faster than AI's capabilities.
- Education's core role in the age of AI is to equip citizens with critical thinking skills to question information, and actively participate in democratic decision-making.

# Diversity and Inclusion

## Diversity within the technology teams

There cannot be a responsible adoption of AI without a serious commitment around diversity and inclusion. First, there should be more diversity within the firms that design and deploy AI across all dimensions from race, gender, age, ethnicity, sexual orientation, educational background, religious views, etc. Research has repeatedly shown that diversity and inclusion lead to innovation, better risk management, and better performance. A 2015 study by McKinsey, for instance, found that there is a linear relationship between teams' diversity and firms' financial returns, confirming that companies with more diverse workforces perform better financially. Such diversity is indispensable to provide different perspectives and life experiences across different areas.

It is key to treat this not as a check-the-box exercise, but to truly look for the best available talent. Not only competent and well trained, but also passionate and committed to having a positive impact. Ensuring that people from diverse groups have a seat at the table is not sufficient, it is also important to create an inclusive environment where everyone has a voice and different views are heard.

Unfortunately, despite the efforts made by most of the biggest players, firms are still behind. For instance, from a racial diversity perspective, a 2019 study by AI Now reported that less than 5% of the US staff at Google, Microsoft, and Facebook are Black, while the Black population represents about 12% of the workforce in the US. As a result of this disparity, there is a much higher risk that all those products used by billions of users may propagate in a very large scale the bias (or lack of awareness) present in those teams. Addressing the lack of a strong and diverse talent pipeline in the technology space will require robust action plans from both firms and national governments aimed at increasing digital literacy and fluency but also strategically attracting the best global talent through scholarships and immigration policies.

In regard to gender diversity, the same study showed that women make up only 15% of AI researchers and engineers at Facebook and just 10% at Google. Not much encouragement comes from the data on women's enrollment in computer science programs at US universities, as it is dramatically down to just 18%, when compared to 37% in 1984. The same research also shows that in addition to being a minority in those contexts, women are less likely to be given the same authority as men in comparable positions.

The lack of female representation has some tangible effects. It is interesting, for example, to note that most names and voices of virtual assistants (Alexa, Siri, Cortana, etc.) are females. This polarization has been highly criticized as reinforcing gender bias and encouraging the stereotype of women as submissive and compliant.

Diversity is needed not only among the business and technical teams, but also among the workers tasked with transactional but very ethically relevant work, such as the annotation of pictures used then to train AI. If those workers lack diversity, the labeling of the images will inevitably reflect their bias. For example, the workers labeling pictures of wedding dresses might end up labeling a colorful North Indian bride as wearing a costume rather than a wedding dress. As those tasks are often outsourced to offshore firms, it might be very difficult to be aware of the levels of diversity and quickly identify related risks, unless there are specific programs to require diversity within the third-party firms under contract.

### Inclusiveness of the models and their data

Diversity checklists/tollgates should also become a required step in the development process of new solutions, to ensure that the data used to run models is representative of all groups. AI models are only as good as the data they are fed with. If the data is skewed as per the view of the person building it, it can have negative consequences on certain groups. As we have seen on many occasions, certain population groups are often under or overrepresented in the training sets. This may result in higher error rates penalizing those groups. We saw, for example, how miserably face recognition tools have failed due to the lack of diversity in the training data with regard to specific groups such as females with darker skin.

### Accessibility of the technology

As technologies are designed, proactive steps need to be taken to ensure the inclusion of people with disabilities and vulnerable citizens. For instance, there are AI solutions built using facial recognition technology that might work well for most people, but not at all for certain groups such as people with different facial characteristics due to injuries or medical conditions. An AI application used in job interviews to assess candidates based on facial features and non-verbal behavior, for instance, might negatively affect candidates unable to control body movements, facial expressions, or eye movements due to hidden disabilities.

Developing inclusive solutions also includes addressing instances in which groups might be excluded due to challenges related to access to the digital world. For example, we might take for granted that a job applicant will be able to submit a resume in digital format, but it might be quite difficult in certain areas to have access to computers and internet connections or access to related training. Similar forms of exclusion emerged during the COVID-19 pandemic: the fact that most of the scheduling of vaccine appointments required online reservations made it very difficult for elderly people and those living in remote areas to have access to the vaccine.

## Inclusion of stakeholders

More in general, there must be a high level of inclusion of those that may be impacted by a new technology. As Aristotle said, it is good to listen to those who are closer to the issue you are investigating even if they are not experts. The people living in the house know the house better than the architect who built it. Opportunities to increase inclusiveness exist across the technology life cycle, from the initial ideation to the design phase, its deployment, and the ongoing monitoring of its performance and impact. When it comes to the future of robotics and AI, everyone is a stakeholder and, therefore, everyone should be included, starting with those that are most exposed to the risks. For example, when developing AI-driven sentencing software, in addition to involving end-users such as the people working in the criminal justice system, we need to also include all the citizens whose lives might be affected by the deployment of such technologies.

## AI as part of the solution

AI could play a role driving more diversity and inclusion and become part of the solution. For example, there are now AI-driven tools that help firms assess the organizational diversity across all groups using deep neural networks. This technology can help minimize the risk of the human factor and provide a set of real-time unbiased reports to all stakeholders with insights for new campaigns to further include diversity.

AI tools can also help address unconscious bias. For instance, models can review job postings and detect language that might be reinforcing bias and help change it to be more appealing to a broader range of candidates. Similar analysis can be done to identify potentially unfair HR practices by comparing employees career advancement and compensation.

## CHAPTER'S KEY POINTS

- Diversity and inclusion across all dimensions of diversity are essential to ensure an ethical deployment of AI.
- The accessibility of the technology and the inclusion of all affected stakeholders throughout its lifecycle should be a key priority for firms deploying new AI-driven solutions.
- AI can itself help identify biases and gaps within organizations and help reach diversity and inclusion targets.

# Reformulating our Priorities

Innovation comes with new and exciting things, and it is easy to be carried away by so many opportunities. However, as we continue to develop new technologies, we should always remember that the ultimate objective, our true North Star, is not innovation itself, but what is beneficial for mankind and improves the human experience. The same principle is often referred to in Western cultures as the pursuit of happiness. The concept has been around for thousands of years, from Aristotle to John Locke, and it has been included in international treaties and national constitutions. One of the most recognized statements of this idea is included in the 1776 US Declaration of Independence, where Life, Liberty, and the Pursuit of Happiness are described as unalienable rights. This idea goes back at least 2,300 years, all the way to Aristotle's principles captured in his Book of Ethics. In his book, Aristotle advocated for the concept of εὐδαιμονία (ev-de-mo-neea): the pursuit of happiness intended as the flourishing and fulfilling of our potential. It is our responsibility to ensure that our civilization will use AI's power so that it will in fact become the best thing that has ever happened to us rather than the opposite.

With so many urgent planetary problems that require action, why should we focus on AI? When we think about the hundreds of millions of people who today do not have water, food, and shelter, or perhaps of the countless human rights violations happening every day or all the environmental problems we are causing, it seems a bit out of touch to worry about AI. Focusing on AI, however, is key because, as we have seen, AI tends to be a multiplier of everything it touches with the potential of being both the best and the worst thing that has ever happened to us. Unless proper controls are built, the deployment of AI will inevitably cause more violence, poverty, inequality, and loss of biodiversity. At the same time, if we find a way to manage it properly, AI could be the solution for many of those problems and even help us mitigate the most critical existential risks such as pandemics or climate change so that our civilization has the right environment to continue to evolve and prosper. There is a great deal of anxiety toward robots and AI, due to the fear that they will take our jobs, manipulate us, and then finally take over the world. However, AI might very well be an amazing opportunity to create a successful synergy between biological and non-biological matter and allow us to be better positioned for the next challenges we will face as humanity.

Even if we all agree that AI ethics should be a priority, how can we use it to best fulfill our potential? Going back to Greek philosophy, Plato provided invaluable and still very relevant guidance on how to do that. In his book called Protagoras, he described the maxims inscribed on the columns standing by the entrance of the Temple of Apollo in Delphi: "Know thyself" and "Nothing in excess". Why are they relevant? "Know thyself" is essential because to unlock AI's most beneficial potentials we need to learn more about ourselves, as individuals and as a civilization. It is up to us to identify the values that matter the most and decide together what future we want to build. To choose wisely, we need to reflect on who we are and what society we want. It is not just about prioritizing certain types of technological solutions, but also about prioritizing across all the (mostly self-inflicted) global challenges we are facing such as poverty, inequality, environmental issues, military conflicts, nuclear weapons proliferation, human rights violations, etc. "Nothing in excess" is also key because to avoid catastrophic disasters in the process, we also need to be humble and recognize our limits. So many failures happened because of incompetence, arrogance, and recklessness. Today's biggest risks do not come from the machines themselves but from the people that build, control, and use them, and from the action or inaction of government and other institutions.

If we do this right, we will experience a rebirth of philosophy. In a world in which activities that require intelligence are no longer an exclusive prerogative of living organisms, we will need to redefine fundamental terms such as intelligence, creativity, and agency. We will also need to reassess our strengths, our weaknesses, what we believe makes us unique, and our role in the world. Even more importantly, we will need to find a way to reassess our values, our priorities, and ultimately determine what society we want to live in, so that we can channel AI's power to help us get closer to it and set up the right type of governance to ensure it does what we want it to do and it never does what we do not want it to do so that AI makes life safer, more efficient, and more humane, by taking on dangerous, boring, or repetitive tasks while preserving areas where human judgment, creativity, and dignity matter most. Philosophy will be key in this process because it will help us better understand the foundations of our values, question the status quo, uncover contradictions and unfounded beliefs, assess our options, and evaluate their tradeoffs. The type of philosophy that we need will be less about conceptual games or debating philosophical models and more about understanding how to best use technology so that it is truly beneficial.

Given AI's power to amplify the impact of what we do, our moral accountability is greater than it has ever been, and we need to be very intentional about setting our priorities. For example, given the scarcity of our resources we need to take the time to collectively take a sober look at our long-term goals and ensure the resources are used wisely and not wasted for what we do not need. Just by looking at the permanent environmental harm that is caused by what it takes to build and power machines (mining metals, building chips, generating electricity, etc.), it is evident how difficult it is going to be. It is in theory possible that there might be situations in which permanent damages to our communities and our ecosystem come with advantages that outweigh the negatives, but that analysis needs to be part of a risk assessment involving the communities affected by those decisions, rather than something decided behind closed doors by private entities or undemocratic institutions. A lot of amazing work in this area has already been completed and is ready to be turned into action. For example, the United Nations' Sustainable Development Goals, comprising of 17 goals spanning from economic development and climate action to access to water and quality education, were adopted by all members in 2015 with the objective to achieve those goals by 2030. With the right focus, AI might very well turn out to be an accelerating factor to help us reach those goals because it can help us collaborate, obtain better data, become more efficient, make better decisions, and build a better world.

It is impossible to predict how many years we have in front of us as a civilization. What we know is that life on Earth appeared 3.5 billion years ago, and that our species has been around for about 100,000 years. If we are, let's say, in the middle of our journey, we have another 100,000 years of potentially mind-blowing adventures. Depending on how well we address our biggest challenges, we might set up the foundation for an incredible future and help humanity flourish like never before.

Fasten your seatbelt. We are entering turbulent times. Technology gives us the power to make profound changes to our world. This could be a new golden age of humanity, but we need to get it right. Together, we need to create the right conditions to ensure this new technology will be favorable to us. A scenario where AI is beneficial is very much possible, but it will not happen automatically.

# Bibliography and Notes

Andy Luc, "Critical Analysis on Deep Learning", Medium, June 6, 2019, https://medium.com/@rokaandy/critical-analysis-on-deep-learning-e13f960c98e

"Google developing kill switch for AI", BBC News, June 8 2016, https://www.bbc.com/news/technology-36472140

Eda Kavlakoglu, "AI vs. Machine Learning vs. Deep Learning vs. Neural Networks: What's the Difference", IBM, May 27 2020, https://www.ibm.com/cloud/blog/ai-vs-machine-learning-vs-deep-learning-vs-neural-networks

Anil Ananthaswamy, "Researchers Build AI That Builds AI", Quantamagazine, January 25 2022, https://www.quantamagazine.org/researchers-build-ai-that-builds-ai-20220125/

Brian Fung, "Investigators have released photos of that deadly Tesla autopilot crash", The Washington Post, July 26 2016, https://www.washingtonpost.com/news/the-switch/wp/2016/07/26/investigators-have-released-photos-and-more-details-on-that-deadly-tesla-crash/

Steven Johnson, "AI Is Mastering Language. Should We Trust What It Says?" The New York Times, April 15 2022, https://www.nytimes.com/2022/04/15/magazine/ai-language.html?login=email&auth=login-email

For more information see "On the Dangers of Stochastic Parrots: Can Language Models Be Too Big?", FAccT '21: Proceedings of the 2021 ACM Conference on Fairness, Accountability, and TransparencyMarch 2021 Pages 610–623https://doi.org/10.1145/3442188.3445922

Eduardo Reck Miranda etc., "An approach to interfacing the brain with quantum computers: practical steps and caveats", January 4 2022, https://arxiv.org/abs/2201.00817

https://arxiv.org/abs/2201.00817

Meredith Broussard, "Artificial Unintelligence, How Computers Misunderstand the World", The MIT Press, April 2018

"Chinese AI caught out by face in bus ad", BCC News, November 27 2018, https://www.bbc.com/news/technology-46357004

Akhila Narla, Brett Kuprel, Kavita Sarin, Roberto Novoa, Justin Ko, "Automated Classification of Skin Lesions: From Pixels to Practice", Journal of Investigative Dermatology, Volume 138, Issue 10, 2018, https://www.sciencedirect.com/science/article/pii/S0022202X18322930/

Max Tegmark, Life 3.0: Being Human in the Age of Artificial Intelligence, Knopf, August 2017

"The Chinese Room Argument", Stanford Encyclopedia of Philosophy, March 19 2004, https://plato.stanford.edu/entries/chinese-room/

Nitasha Tiku, "The Google engineer who thinks the company's AI has come to life", The Washington Post, June 11 2022, https://www.washingtonpost.com/technology/2022/06/11/google-ai-lamda-blake-lemoine/

https://www.nature.com/articles/s41589-023-01349-8.epdf?sharing_token=RWd0NhirFpy2b2BGPZP5v9RgN0jAjWel9jnR3ZoTv0PgqDK_eQp8NcAST1F90jpq3WEGXG2Su3RGayoL_4HVm-885WktBHURHlCxulwlcvpyGtITV-DP3D206obw-VwWHxYxnDYB7coCRf0l466lVvYqSrbMGRSnrbQLgtJwzb9l1g8rqpxmCL-4_-k8FciBA9RvvjphptrsdaCOzV6I g%3D%3D&tracking_referrer=www.bbc.com

"Japanese cafe uses robots controlled by paralysed people", BBC News, https://www.bbc.com/news/technology-46466531

Ewen Callaway, "DeepMind's AI makes gigantic leap in solving protein structures", Nature, November 30 2020, https://www.nature.com/articles/d41586-020-03348-4

Marshall Allen, "Health Insurers Are Vacuuming Up Details About You , And It Could Raise Your Rates

Darrell M. West and John R .Allen, "Turning Point", Brookings Institution Press, 2020

Renee Cho, "Artificial Intelligence, A Game Changer for Climate Change and the Environment", Columbia Climate School, June 5 2018, https://news.climate.columbia.edu/2018/06/05/artificial-intelligence-climate-environment/

Degrave, J., Felici, F., Buchli, J. et al. "Magnetic control of tokamak plasmas through deep reinforcement learning", Nature 602, 414–419 (2022), https://doi.org/10.1038/s41586-021-04301-9

Paul Gillin, "Tackling Data Center Water Usage Challenges Amid Historic Droughts, Wildfires", Data Center Frontier, January 8 2021, https://www.datacenterfrontier.com/special-reports/article/11428474/tackling-data-center-water-usage-challenges-amid-historic-droughts-wildfires

"Exposed: Child labour behind smart phone and electric car batteries", Amnesty.org, January 19 2016, https://www.amnesty.org/en/latest/news/2016/01/child-labour-behind-smart-phone-and-electric-car-batteries/

Steven Pinker, "What Do You Think About Machines That Think?", Edge.Org, https://www.edge.org/response-detail/26243

Tim Harford, "What an abusive chatbot can teach us about the art of conversation", Financial Times, March 17 2022, https://www.ft.com/content/9cf3717a-2023-4564-a953-7973dcc14a26

Barry Collins, "I Interviewed ChatGPT About AI Ethics , And It Lied To Me", Forbes, December 2, 2022, https://www.forbes.com/sites/barrycollins/2022/12/02/i-interviewed-an-ai-about-the-ethics-of-aiand-it-lied-to-me/?sh=22e109276253

Cade Metz, "Who Is Making Sure the A.I. Machines Aren't Racist?", The New York Times, March 15 2021, https://www.nytimes.com/2021/03/15/technology/artificial-intelligence-google-bias.html

Phil La Duke, "Robots Are Stealing Our Jobs", Entrepreneur, April 23 2019, https://www.entrepreneur.com/science-technology/robots-are-stealing-our-jobs/332468

Sayash Kapoor, Arvind Narayanan, "Eighteen pitfalls to beware of in AI journalism", Sept 30 2022, https://aisnakeoil.substack.com/p/eighteen-pitfalls-to-beware-of-in

Matthew Jarrett, "Soviet Child Mines", forgottenhistory.me, May 31 2022, https://www.forgottenhistory.me/new-blog/soviet-child-mines

Ben Wagner, "Ethics As An Escape From Regulation. From "Ethics-Washing" To Ethics-Shopping?", Amsterdam University Press, 2018

World Health Organization (WHO) "Global Status Report on Road Safety", December 2018, https://www.who.int/violence_injury_prevention/road_safety_status/2018/en/external icon

Adrienne LaFrance, "Self-Driving Cars Could Save 300,000 Lives Per Decade in America", The Atlantic, September 29 2015, www.theatlantic.com%2Ftechnology%2Farchive%2F2015%2F09%2Fself-driving-cars-could-save-300000-lives-per-decade-in-america%2F407956%2F&usg=AOvVaw0jKrI67L7SWEN_SwT52oTB

https://www.moralmachine.net

Adrienne Mayor, Gods and Robots: Myths, Machines, and Ancient Dreams of Technology, Princeton University Press, April 21 2020

Ernest Davis and Gary Marcus, Rebooting AI, Pantheon, September 2019

Hailey Austine Scherer, "Metaphors Matter: Top-Down Effects on Anthropomorphism", Dartmouth College, May 28 2020, http://digitalcommons.dartmouth.edu/senior_theses/270

Küster, D., Swiderska, A., & Gunkel, D. (2021). "I saw it on YouTube! How online videos shape perceptions of mind, morality, and fears about robots", New Media & Society, 23(11), 3312–3331. https://doi.org/10.1177/14614448209541999

Samuel Gibbs, "Hackers can hijack Wi-Fi Hello Barbie to spy on your children", The Guardian, November 26 2015, https://www.theguardian.com/technology/2015/nov/26/hackers-can-hijack-wi-fi-hello-barbie-to-spy-on-your-children

Mark Coeckelbergh, Robot Ethics, MIT Press, 2022

Homer, Iliad 18. 416 ff (Greek epic C8th B.C.)

Joanna J. Bryson, Robots Should Be Slaves, May 21 2009, http://www.cs.bath.ac.uk/~jjb/ftp/Bryson-Slaves-Book09.pdf

Adrienne Mayor, Gods and Robots: Myths, Machines, and Ancient Dreams of Technology, Princeton University Press, April 21 2020

Slaughterbots, https://www.youtube.com/watch?v=HipTO_7mUOww

Thomas Nilsen, "World's longest nuclear submarine handed over to the Russian Navy", The Barents Observer, July 8 2022, https://thebarentsobserver.com/en/security/2022/07/worlds-longest-nuclear-submarine-handed-over-russian-navy

Davis Hambling, "Israel used world's first AI-guided combat drone swarm in Gaza attacks", New Scientist, June 30 202, https://www.newscientist.com/article/2282656-israel-used-worlds-first-ai-guided-combat-drone-swarm-in-gaza-attacks/

Official Document System of the United Nations, https://documents-dds-ny.un.org/doc/UNDOC/GEN/N21/037/72/PDF/N2103772.pdf?OpenElement

https://www.theverge.com/2022/11/30/23486090/robots-deadly-force-kill-san-francisco-police-approve

https://autonomousweapons.org

https://towardsdatascience.com/a-new-spin-to-ethical-ai-trolley-problems-with-gpt-3-b7462be81899

https://openai.com/blog/chatgpt/

Abeba Birhane, Deborah Raji, "ChatGPT, Galactica, and the Progress Trap", Wired, Dec 9, 2022, https://www.wired.com/story/large-language-models-critique/

Martin Coulter and Greg Bensinger, "Alphabet shares dive after Google AI chatbot Bard flubs answer in ad", Reuters, February 8. 2023, https://www.reuters.com/technology/google-ai-chatbot-bard-offers-inaccurate-information-company-ad-2023-02-08/

Ian Bogost, "ChatGPT Is Dumber Than You Think", The Atlantic, December 7, 2022, https://www.theatlantic.com/technology/archive/2022/12/chatgpt-openai-artificial-intelligence-writing-ethics/672386/

Dina Bass, "Microsoft escalates talks with the chatbot threatening Google's dominance", Fortune, January 10, 2023, https://fortune.com/2023/01/10/microsoft-eyeing-10-billion-investment-openai-chatgpt-chatbot-threatening-google-dominance/

Catherine Stupp, "Fraudsters Used AI to Mimic CEO's Voice in Unusual Cybercrime Case", The Wall Street Journal, August 30, 2019, https://www.wsj.com/articles/fraudsters-use-ai-to-mimic-ceos-voice-in-unusual-cybercrime-case-11567157402

https://www.vice.com/en/article/pkadgm/man-dies-by-suicide-after-talking-with-ai-chatbot-widow-says

Matthew Hutson, "Robo-writers: the rise and risks of language-generating AI", Nature, March 3 2021, https://www.nature.com/articles/d41586-021-00530-0

Kevin Roose, "The Brilliance and Weirdness of ChatGPT", The New York Times, December 5, 2022, https://www.nytimes.com/2022/12/05/technology/chatgpt-ai-twitter.html?searchResultPosition=1

Leigh Alexander, "Do Google's 'unprofessional hair' results show it is racist?", The Guardian, April 8 2016, https://www.theguardian.com/technology/2016/apr/08/does-google-unprofessional-hair-results-prove-algorithms-racist-

Chris Sherman, "Trolls propel Trump to top Google ratings for idiot, July 19, 2018 , https://searchengineland.com/trolls-game-google-to-get-trump-in-top-results-for-idiot-302374

Ellora Thadaney Israni, "When an Algorithm Helps Send You to Prison", The New York Times, October 26 2017, https://www.nytimes.com/2017/10/26/opinion/algorithm-compas-sentencing-bias.html

Obermeyer, Ziad Powers, Brian Vogeli, Christine et al. "Dissecting racial bias in an algorithm used to manage the health of populations", Science, October 1 2019, https://escholarship.org/content/qt6h92v832/qt6h92v832.pdf

Theodore Kim, "AI flaws could make your next car racist", Los Angeles Times, October 7 2021, https://www-latimes-com.cdn.ampproject.org/c/s/www.latimes.com/opinion/story/2021-10-07/op-ed-ai-flaws-could-make-your-next-car-racist?_amp=true

Jana Kasperkevic "Google says sorry for racist auto-tag in photo app", The Guardian, July 1 2015, https://www.theguardian.com/technology/2015/jul/01/google-sorry-racist-auto-tag-photo-app

Virginia Eubanks, Automating Inequality: How High-Tech Tools Profile, Police, and Punish the Poor, St. Martin's Press, January 2018

Harwell, "Facial recognition firm Clearview AI tells investors it's seeking massive expansion beyond law enforcement", The Washington Post, February 16 2022, ttps://www.washingtonpost.com/techno logy/2022/02/16/clearview-expansion-facial-recognition/

Alex Scroxton, "Mass health tracker data breach has UK impact" , Computer Weekly, September 12 2021, https://www.computerweekly.com/news/252506664/Mass-health-tracker-data-breach-has- UK-impact

Geoffrey A. Fowler, "Alexa has been eavesdropping on you this whole time", The Washington Post, May 6 2019, https://www.washingtonpost.com/technology/2019/05/06/alexa-has-been-eavesdropping-you-this-whole-time/

Jeb Su, "Apple Apologizes For Eavesdropping On Customers, Keeping Siri Recordings Without Permission" Forbes, August 28 2019, https://www.forbes.com/sites/jeanbaptiste/2019/08/28/apple-apologizes-for-eavesdropping-on-customers-keeping-siri-recordings-without-permission/?sh=6236afe37b8f

https://www.cbsnews.com/news/tesla-lawsuit-privacy-concerns-autopilot-reuters/?ftag=CNM-00-10aac3a

Liza Lin, "A World With a Billion Cameras Watching You Is Just Around the Corner, The Wall Street Journal", December 6 2019, https://www.wsj.com/articles/a-billion-surveillance-cameras-forecast-to-be-watching-within-two-years-11575565402

Baltimore Police Secretly Running Aerial Mass-Surveillance Eye in the Sky, https://www.aclu.org/news/privacy-technology/baltimore-police-secretly-running-aerial-mass

Yuriy Gorodnichenko, Tho Pham and Oleksandr Talavera. "Social media, sentiment and public opinions: evidence from #brexit and #uselection" NBER Working Paper 24631. May 2018, https://www.nber.org/papers/w24631.pdf

Carole Cadwalladr and Emma Graham-Harrison, "How Cambridge Analytica turned Facebook 'likes' into a lucrative political tool", The Guardian, March 17 2018, https://www.theguardian.com/technology/2018/mar/17/facebook-cambridge-analytica-kogan-data-algorithm.

New York university, "Social media use driven by search for reward, akin to animals seeking food", Science Daily, February 26 2021, https://www.sciencedaily.com/releases/2021/02/210226103811.htm

Elle Hunt, "Tay, Microsoft's AI chatbot, gets a crash course in racism from Twitter", The Guardian, March 24 2016, https://www.theguardian.com/technology/2016/mar/24/tay-microsofts-ai-chatbot-gets-a-crash-course-in-racism-from-twitter

Act to add Chapter 6 (commencing with Section 17940) to Part 3 of Division 7 of the Business and Professions Code, relating to bots. S.1001, California State Senate, September 28 2018, https://leginfo.legislature.ca.gov/faces/billTextClient.xhtml?billid=201720180SB1001

https://www.youtube.com/watch?v=OWihqvqSMAQ

Andrew Liptak, "Amazon's Alexa started ordering people dollhouses after hearing its name on TV", The Verge, Jan 7 2017, https://www.theverge.com/2017/1/7/14200210/amazon-alexa-tech-news-anchor-order-dollhouse

Hamza Shaban, "An Amazon Echo recorded a family's conversation, then sent it to a random person in their contacts, report says", The Washington Post, May 24 2018, https://www.washingtonpost.com/news/the-switch/wp/2018/05/24/an-amazon-echo-recorded-a-familys-conversation-then-sent-it-to-a-random-person-in-their-contacts-report-says/

Hope King, "Woman's home demolished after Google Maps error", CNN, March 25 2016, https://money.cnn.com/2016/03/25/technology/google-maps-house/

"Knight Capital lost about US$10M a minute as a result of trading disaster, Financial Post, August 2, 2012, https://financialpost.com/investing/knight-capital-losing-about-us10m-a-minute-after-trading-disaster

Alfred Ng, "300-pound security robot goes haywire, runs over 1-year-old boy at California mall", New York Daily News, July 12 2016, https://www.nydailynews.com/news/national/security-robot-haywire-runs-boy-1-calif-mall-article-1.2708975

John Henley, “Chess robot grabs and breaks finger of seven-year-old opponent”, The Guardian, July 23 2022, https://www.theguardian.com/sport/2022/jul/24/chess-robot-grabs-and-breaks-finger-of-seven-year-old-opponent-moscow?CMP=share_btn_link

https://arstechnica-com.cdn.ampproject.org/c/s/arstechnica.com/tech-policy/2023/01/facial-recognition-error-led-to-wrongful-arrest-of-black-man-report-says/amp/

Pranshu Verma and Will Oremus, “ChatGPT invented a sexual harassment scandal and named a real law prof as the accused”, the Washington Post, April 5 2023, https://www.washingtonpost.com/technology/2023/04/05/chatgpt-lies/

Dan Milmo, “Two US lawyers fined for submitting fake court citations from ChatGPT”, The Guardian, June 23, 2023, https://www.theguardian.com/technology/2023/jun/23/two-us-lawyers-fined-submitting-fake-court-citations-chatgpt

Katie Spalding, “This Viral List Shows Some Of The Most Hilarious Ways AIs Have Tried To Outsmart Humans”, IFLScience, November 9, 2016, https://www.iflscience.com/this-viral-list-shows-some-of-the-most-hilarious-ways-ais-have-tried-to-outsmart-humans-50513

Robert Wiblin and Keiran Harris,
“The flaws that make today’s AI architecture unsafe and a new approach that could fix it”, https://80000hours.org, June 22 2022, https://80000hours.org/podcast/episodes/stuart-russell-human-compatible-ai/

Carl Benedikt Frey and Michael A. Osborne, “The future of employment: how susceptible are jobs to computerisation?”, September 17 2013, https://www.oxfordmartin.ox.ac.uk/downloads/academic/The_Future_of_Employment.pdf

International Federation of Robotics, “Robot Density nearly Doubled globally”, December 14 2021, https://ifr.org/ifr-press-releases/news/robot-density-nearly-doubled-globally

Will Evans, “Ruthless quotas at Amazon are maiming employees”, The Atlantic, November 25, 2019, https://www.theatlantic.com/technology/archive/2019/11/amazon-warehouse-reports-show-worker-injuries/602530/

Oscar Williams-Grut, “Nobel economist: 'i don't think globalization is anywhere near the threat that robots are”, Business Insider, December 24, 2016, https://www.businessinsider.in/nobel-economist-i-dont-think-globalisation-is-anywhere-near-the-threat-that-robots-are/articleshow/56153603.cms

Luc Olinga, “Everything Elon Musk Said At Tesla AI Day to Sell His Vision”, The Street, October 1 2022, https://www.thestreet.com/technology/everything-elon-musk-said-at-tesla-ai-day-to-sell-his-vision

Jeffrey D. Sachs And Laurence J. Kotlikoff, “Mart Machines And Long-Term Misery”, National Bureau Of Economic Research, December 2012, https://www.nber.org/system/files/working_papers/w18629/revisions/w18629.rev0.pdf

Daron Acemoglu and Pascual Restrepo, “Robots and Jobs: Evidence from US labor markets”, October 2016, https://pdfs.semanticscholar.org/70d4/05c8d11d042f7947828bbaae9ec0a5ed003f.pdf

Paola Tubaro, Antonio A. Casilli, “Portraits of c-workers: The real people behind AI in France”, 2nd Crowdworking Symposium 2020, Research program “Digital Future”, Universities of Paderborn and Bielefeld, Oct 2020, Paderborn, Germany. hal-02960775

Dave Lee, “Why Big Tech pays poor Kenyans to teach self-driving cars”, BBC, November 3, 2018, https://www.bbc.com/news/technology-46055595

Billy Perrigo, "OpenAI Used Kenyan Workers on Less Than $2 Per Hour to Make ChatGPT Less Toxic", Time, January 18 2023, https://time.com/6247678/openai-chatgpt-kenya-workers/

Ryan Browne, "Elon Musk warns A.I. could create an immortal dictator from which we can never escape", CNBC, April 6 2018, https://www.cnbc.com/2018/04/06/elon-musk-warns-ai-could-create-immortal-dictator-in-documentary.html

Frank Pasquale, "New Laws of Robotics", The Belknap Press, 2020

Kimberly Ricci, "Report: Uber And Arizona's Governor Kept The Self-Driving Car Program A Secret", Uproxx, March 28 2018, https://uproxx.com/culture/uber-arizona-governor-self-driving-secret-fatal-crash/

Javier E. David, 'Uber hammered by price gouging accusations during NYC's explosion", CNBC, September 18 2016, https://www.cnbc.com/2016/09/18/uber-hammered-by-price-gouging-accusations-during-nycs-explosion.html

Bobbie Johnson and Gideon, "He Hey Google, sorry you lost your ethics council, so we made one for you", MIT Technology Review, April 6 2019, https://www.technologyreview.com/2019/04/06/65905/google-cancels-ateac-ai-ethics-council-what-next/

https://www.youtube.com/watch?v=2C-A797y8dA

Brad Smith, "Facial Recognition Technology: The Need for Public Regulation and Corporation Responsibility", Microsoft, June 13 2018, https://blogs.microsoft.com/on-the-issues/2018/07/13/facial-recognition-technology-the-need-for-public-regulation-and-corporate-responsibility/

https://ec.europa.eu/info/law/law-topic/data-protection/data-protection-eu_en

The General Data Protection Regulation (GDPR), the Data Protection Law Enforcement Directive and other rules concerning the protection of personal data, https://digital-strategy.ec.europa.eu/en/library/ethics-guidelines-trustworthy-ai

Luciano Floridi, "Soft ethics and the governance of the digital", Philosophy & Technology, 31(1), 1–8, 2018, https://link.springer.com/article/10.1007/s13347-018-0303-9

Lee Fang, "Google hedges on promise to end controversial involvement in military drone contract", The Intercept, March 1 2019, https://theintercept.com/2019/03/01/google-project-maven-contract/

https://futureoflife.org/2017/08/11/ai-principles/

Dame Vivian Hunt, Dennis Layton, and Sara Prince, "Why Diversity Matter", McKinsey & Company, January 2015, https://www.mckinsey.com/business-functions/people-and-organizational-performance/our-insights/why-diversity-matters

Discriminating Systems, AI Now, April 2019, https://ainowinstitute.org/discriminatingsystems.pdf

Ho Caitlin Chin and Mishaela Robisonw, "How AI bots and voice assistants reinforce gender bias", Brookings, November 23 2020, https://www.brookings.edu/research/how-ai-bots-and-voice-assistants-reinforce-gender-bias/

Sean L. Metzger et all, "A high-performance neuroprosthesis for speech decoding and avatar control", Nature, August 23 2023, https://www.nature.com/articles/s41586-023-06443-4

Alex Hughes, “This robot arm can be controlled by the power of your brain”, BBC Science Focus, January 20, 2022 www.sciencefocus.com/news/this-robot-arm-can-be-controlled-by-the-power-of-your-brain

Luis Prada, “Constant Scrolling Is Giving AI Brain Rot”, Vice, October 24, 2025, www.vice.com/en/article/constant-scrolling-is-giving-ai-brain-rot/

Mathura Sundaram, “The impact of digital technology, social media, and artificial intelligence on cognitive functions: a review”, Frontiers, 23 November 2023, www.frontiersin.org/journals/cognition/articles/10.3389/fcogn.2023.1203077/full

Kim Kyung-Hoon and Satoshi Sugiyama, “AI romance blooms as Japanese woman weds virtual partner of her dreams”, Reuters, November 2025, www.reuters.com/investigates/special-report/japan-ai-wedding/

For permission requests or enquiries, please contact:
robethics@gmail.com

Interested in getting involved?
Join the Responsible AI community at robethics.org

www.ingramcontent.com/pod-product-compliance
Lightning Source LLC
LaVergne TN
LVHW050757061126
851857LV00012B/595
*9798234199423*